SpringerBriefs in Electrical and Computer Engineering

Speech Technology

Series Editor

Amy Neustein

For further volumes:
http://www.springer.com/series/10043

Editor's Note

The authors of this series have been hand selected. They comprise some of the most outstanding scientists—drawn from academia and private industry—whose research is marked by its novelty, applicability, and practicality in providing broad-based speech solutions. The Springer Briefs in Speech Technology series provides the latest findings in speech technology gleaned from comprehensive literature reviews and empirical investigations that are performed in both laboratory and real life settings. Some of the topics covered in this series include the presentation of real life commercial deployment of spoken dialog systems, contemporary methods of speech parameterization, developments in information security for automated speech, forensic speaker recognition, use of sophisticated speech analytics in call centers, and an exploration of new methods of soft computing for improving human–computer interaction. Those in academia, the private sector, the self service industry, law enforcement, and government intelligence are among the principal audience for this series, which is designed to serve as an important and essential reference guide for speech developers, system designers, speech engineers, linguists, and others. In particular, a major audience of readers will consist of researchers and technical experts in the automated call center industry where speech processing is a key component to the functioning of customer care contact centers.

Amy Neustein, Ph.D., serves as editor in chief of the International Journal of Speech Technology (Springer). She edited the recently published book Advances in Speech Recognition: Mobile Environments, Call Centers and Clinics (Springer 2010), and serves as quest columnist on speech processing for Womensnews. Dr. Neustein is the founder and CEO of Linguistic Technology Systems, a NJ-based think tank for intelligent design of advanced natural language-based emotion detection software to improve human response in monitoring recorded conversations of terror suspects and helpline calls.

Dr. Neustein's work appears in the peer review literature and in industry and mass media publications. Her academic books, which cover a range of political, social, and legal topics, have been cited in the Chronicles of Higher Education and have won her a pro Humanitate Literary Award. She serves on the visiting faculty of the National Judicial College and as a plenary speaker at conferences in artificial intelligence and computing. Dr. Neustein is a member of MIR (machine intelligence research) Labs, which does advanced work in computer technology to assist underdeveloped countries in improving their ability to cope with famine, disease/illness, and political and social affliction. She is a founding member of the New York City Speech Processing Consortium, a newly formed group of NY-based companies, publishing houses, and researchers dedicated to advancing speech technology research and development.

K. Sreenivasa Rao · Shashidhar G. Koolagudi

Robust Emotion Recognition using Spectral and Prosodic Features

 Springer

K. Sreenivasa Rao
Shashidhar G. Koolagudi
School of Information Technology
Indian Institute of Technology Kharagpur
Kharagpur, West Bengal
India

ISSN 2191-737X
ISSN 2191-7388 (electronic)
ISBN 978-1-4614-6359-7
ISBN 978-1-4614-6360-3 (eBook)
DOI 10.1007/978-1-4614-6360-3
Springer New York Heidelberg Dordrecht London

Library of Congress Control Number: 2012954864

Printed on acid-free paper

Springer is part of Springer Science+Business Media (www.springer.com)

Preface

Human beings use speech as a primary mode of communication for conveying messages. A speech signal carries multiple cues related to intended message, speaker and language identities, behavioural and emotional mood of the speaker and characteristics of background environment. Human beings exploit all these cues for performing various speech tasks. Now a days in all areas of applications, machines are extensively used for performing the desired tasks automatically. For interacting with machines, humans feel speech interface to be the most convenient one. Speech interface to machine should take care of basic tasks such as speech understanding and speech generation.

Most of the present speech systems support the basic speech tasks for the neutral speech recorded in a clean environment. But, in general human speech is mostly embedded with emotions to convey the intended message. Processing emotional speech by a machine is a complex task. In real life, emotions are highly dynamic and depend on several factors such as speaker, language, culture, society and geographical regions. The basic objective of this book is to explore various features and models for characterising and discriminating the emotions. In addition to conventional speech features, emotion-specific spectral and prosodic features are introduced in this book for performing robust emotion recognition.

This book primarily focuses on spectral and prosodic features to discriminate emotions. In addition to conventional spectral features extracted from the entire speech through block processing, sub-syllabic regions such as consonants, vowels and CV (Consonant-to-Vowel) transition regions are explored for computing spectral features. Spectral features extracted through pitch synchronous analysis are explored for reducing the ambiguity in confusable emotions. Prosodic features extracted from words and syllables present at different positions (initial, middle and final) are investigated to analyse the emotion-specific knowledge. Various nonlinear models, such as auto-associative neural networks, support vector machines and Gaussian mixture models, are explored to capture the emotion-specific information from the features mentioned above. In this book, emotional database Indian Institute of Technology Kharagpur Simulated Emotion Speech Corpus (IITKGP-SESC) in an Indian language Telugu, Berlin emotional speech database (Emo-DB) and Hindi movie database are used for analysing the emotion recognition performance.

This book is mainly intended for researchers working on emotion recognition from speech. The book is also useful for young researchers, who want to pursue research in speech processing using basic excitation source, vocal tract and prosodic features. Hence, this may be recommended as the text or reference book at the postgraduate level advanced speech processing course. The book has been organised as follows:

Chapter 1 introduces automatic emotion recognition from speech as one of the thrust areas of research in speech processing. Psychological and engineering aspects of emotional speech signal have been discussed. Chapter 2 discusses the non-conventional spectral features extracted from pitch synchronous analysis and sub-syllabic regions for discriminating the emotions. Chapter 3 discusses the global and local prosodic features extracted from words and syllables for classifying the emotions. Chapter 4 exploits the complementary information provided by various features by combining their evidences using appropriate fusion techniques. Chapter 5 introduces multi-stage emotion classification using speaking rate features. Chapter 6 performs real-life emotion classification using the proposed robust features. Chapter 7 summarises the contents of the book, highlights the contributions of the chapters and discusses the scope for future work.

Many people have helped us during the course of preparation of this book. We would especially like to thank all Professors of School of Information and Technology, IIT Kharagpur, for their moral encouragement and technical discussions during the course of editing and organisation of this book. Special thanks to our colleagues at Indian Institute of Technology, Kharagpur, India and Graphic Era University, Dehradun, India for their co-operation and coordination to carry out the work. We preferably acknowledge the efforts of Anurag Barthwal, Swati Devliyal, Kritik Sharma, Shan-e.Fatima and Deepika Rastogi in preparing the draft of Chap. 6. We are grateful to our parents and family members for their constant support and encouragement. Finally, we thank all our friends and well-wishers.

K. Sreenivasa Rao
Shashidhar G. Koolagudi

Contents

Acronyms

AANN	Autoassociative neural network
CART	Classification and regression tree
CV	Consonant vowel
DCT	Discrete cosine transform
DFT	Discrete fourier transform
F	Female
FD-PSOLA	Frequency domain pitch synchronous overlap and add
FFNN	Feedforward neural network
GC	Glottal closure
Hi	Hindi
HNM	Harmonic plus noise model
IDFT	Inverse discrete fourier transform
IITM	Indian Institute of Technology Madras
IPCG	Inter Perceptual Center Group
IPO	Institute of Perception Research
ITRANS	Common transliteration code for Indian languages
LP	Linear prediction
LPCs	Linear prediction coefficients
LP-PSOLA	Linear prediction pitch synchronous overlap and add
M	Male
MOS	Mean opinion score
NN	Neural network
OLA	Overlap and add
PF	Phrase final
PI	Phrase initial
PM	Phrase middle
PSOLA	Pitch synchronous overlap and add
RFC	*Rise/fall/connection*
RNN	Recurrent neural network

SOP	Sums-of-product
STRAIGHT	Speech Transformation and Representation Using Adaptive Interpolation of WeiGHTed Spectrum
SVM	Support vector machine
Ta	Tamil
TD-PSOLA	Time domain pitch synchronous overlap and add
Te	Telugu
TTS	Text-to-speech
VOP	Vowel onset point
V/UV/S	Voiced/unvoiced/silenc
WF	Word final
WI	Word initial
WM	Word middle

Chapter 1
Introduction

Abstract This chapter briefly discusses about the importance of analysis of emotions from speech signal. Significance of emotions from psychological and signal processing aspects is discussed. Influence of emotions on the characteristics of speech production system is briefly mentioned. Role of excitation source, vocal tract system and prosodic features is discussed in the context of various speech tasks highlighting the task of recognizing emotions. Different types of emotional speech databases used for carrying out various emotion-specific tasks are briefly discussed. Various applications related to speech emotion recognition are mentioned. Important state-of-the-art issues prevailing in the area of emotional speech processing are discussed at the end of the chapter along with a note on the organization of the book.

1.1 Introduction

Speech is a natural mode of communication among human beings. Speech signal carries multi-dimensional information such as intended message, speaking characteristics of a person, spoken language identity, speaker mood and emotions. Humans can capture and utilize all components of speech effortlessly for carrying out various activities in their day to day life. Speech processing by machines is a tough task. At present, machines are able to process speech in presence of several constraints. Most of the existing speech systems will simply fail while processing speech in presence of noisy and emotional environments. For developing a sophisticated speech interface between human and machine, the machine should be capable to process speech like humans. With a learned knowledge from childhood, humans process speech very effectively. From the speech humans can understand basic message, intention of the speaker, recognise the speaker, identify the language used for encoding the message and emotional content of the message. To have a dialogue in speech mode, one should generate speech by embedding all information components such as message, speaker characteristics, language specific information and desired emotions, and should be

K. S. Rao and S. G. Koolagudi, *Robust Emotion Recognition using Spectral and Prosodic Features*, SpringerBriefs in Speech Technology,
DOI: 10.1007/978-1-4614-6360-3_1, © The Author(s) 2013

able to process the speech for capturing the above information sources present in speech. While interacting with a machine, humans expect the machine to perform above speech tasks.

Humans extensively use emotions to express their intentions through speech [1]. The same textual message would be conveyed with different semantics (meaning) by incorporating appropriate emotions. Humans understand the intended message by perceiving the underlying emotions in addition to phonetic information. Therefore, there is a need to develop speech systems that can process emotions along with the message [2]. The basic goals of emotional speech processing are (a) recognizing the emotions present in speech and (b) synthesizing the desired emotions in speech according to the intended message. From the machine's perspective, understanding speech emotions can be viewed as the classification or discrimination of emotions. Synthesis of emotions can be viewed as incorporating the emotion-specific knowledge during speech synthesis. The emotion-specific knowledge is acquired from the emotion models, designed for capturing the emotion-specific characteristics.

In this book, we are focussing on characterization and discrimination of emotions by a machine using speech signal. Emotion processing is very complex task. It involves semantics at higher level, intention of the message and natural characteristics of a speaker. Expression of emotions depends on several factors such as speaker, gender, language, and social constraints. Most of the existing emotion processing systems explored the full-blown simulated emotions. But, in real life emotions are not always full-blown, and mostly blended emotions. It is very difficult to crisply discriminate the natural emotions. Most of the works have explored the conventional spectral and prosodic features for characterizing the emotions. The extreme emotions such as anger and sadness may be discriminated well by using the conventional spectral and prosodic features, but they may not be effective for discriminating other emotions. For affective discrimination of emotions, in this book we have proposed some new spectral and prosodic features. The proposed features are expected to capture the finer variations in emotion specific spectral and prosodic characteristics. The proposed features are evaluated for discriminating the real life emotions.

1.2 Emotion from Psychological Perspective

Emotions have been studied in several scientific disciplines, such as: Biology (Physiological), Psychology, Speech science, Neuroscience, Psychiatry, Anthropology, Sociology, Communication and so on. Subjects like Business Management and Advertising need extensive use of emotion processing. As a result, distinctive perspectives on the concept of *emotion* have emerged, appropriate to the complexity and variety of the emotions. However, it is important to consider these different perspectives not as competitive but as complementary. In this book, emotions are analyzed in view of psychological, physiological and speech perspectives.

Psychology of emotions can be viewed as a complex experience of consciousness (psychology), bodily sensation (physiology), and behavior (action-speech).

The emotions generally represent a synthesis of subjective experience, expressive behavior, and neurochemical activity. There are more than 300 crisply identified emotions by researchers [3, 4]. However generally, all of them are not experienced in day-to-day life. In this regard, most researchers agree on the principles of *Palette theory* that quotes, *any emotion is the composition of six primary emotions as any color is the combination of 3 primary colors* [5]. Anger, disgust, fear, happiness, sadness and surprise are considered as the primary or basic emotions by most researchers [6]. These are also referred as *archetypal emotions* [7].

In psychology, expression of emotions is considered as a response to stimuli that involves characteristic physiological changes [8, 9]. According to physiology, an emotion is defined as a disruption in the homeostatic baseline [8]. Based on these changes, the properties of emotions can be explained in a three-dimensional space. Essential dimensions of emotional states are captured by the features of activation (arousal; measured as an intensity), affect (valence or pleasure; measured as positive or negative feeling after emotion perception) and power (control; measured as dominance or submissiveness in emotion expression). According to the physiology of emotion production mechanisms, it has been found that the nervous system is stimulated by the expression of high arousal emotions like anger, happiness and fear. This phenomenon causes an increased heart beat rate, higher blood pressure, changes in respiration pattern, greater sub-glottal air pressure in the lungs and dryness of the mouth. The resulting speech is correspondingly louder, faster and characterized with strong high-frequency energy, higher average pitch, and wider pitch range [10]. On the other hand, for the low arousal emotions like sadness, the nervous system is stimulated causing the decrease in heart beat rate, blood pressure, leading to increased salivation, slow and low-pitched speech with little high-frequency energy. Thus, acoustic features such as pitch, energy, timing, voice quality, and articulation of the speech signal highly correlate with the underlying emotions [11]. However, distinguishing emotions without any ambiguity, using any one of the above mentioned dimensions, is a difficult task. For example, both anger and happiness have high activation but they convey different affect (valence or pleasure information). This difference is characterized by using both activation and valence dimensions. Hence, emotion is the complex psychophysiological experience of an individual's state of mind as interacting with biochemical (internal) and environmental (external) influences. In humans, emotion fundamentally involves physiological arousal, expressive behaviors, and conscious experience. Therefore the three aspects of emotions are psychological (what one is thinking), physiological (what one's body is doing), and expressive (how one reacts) in nature [12].

1.3 Emotion from Speech Signal Perspective

Psychological and physiological changes caused due to emotional experience lead to certain actions. Speech is one of the important outcomes of the emotional state of human beings. A speech signal is produced from the contribution of the vocal tract system excited by excitation source signal [13]. Hence, speech specific information may be extracted from vocal tract system and excitation source characteristics. The emotion-specific characteristics of the speech can be attributed to (1) characteristics of the excitation source, (2) the sequence of shapes of the vocal tract system while producing different emotions, (3) supra-segmental characteristics (duration, pitch and energy), and (4) linguistic information.

1.3.1 Speech Production Mechanism

The air pressure created in the trachea (sub-glottal region) gets released into the vocal tract and nasal cavity, in the form of air puffs. The pressure created beneath the vocal folds causes their vibration, known as glottal activity. The air flow is chopped into a sequence of quasi-periodic pulses through vibration of vocal folds. This sequence of impulse-like excitation gets transformed into different frequency components, while passing through the pharynx, oral, and nasal cavities [14]. The vibration of vocal folds acts as the major source of excitation to the vocal tract. The oral cavity behaves as a time varying resonator to enhance certain frequency components. Movements of articulator organs of the vocal and nasal tracts modulate this quasi-periodic sequence of air puffs into speech, that gets radiated through lips. Vibration of vocal folds leads to the production of voiced sound units. Constriction of a vocal tract at different places, due to the movement of articulators, along with the noisy excitation leads to production of unvoiced sound units. Depending upon the position of various articulators and the phenomenon of voicing or unvoicing, different sound units are produced. The characteristics of glottal activity and vocal tract shapes also play a major role in modulating different emotions, during production of speech.

1.3.2 Source Features

During production of speech, vibration of vocal folds provides quasi periodic impulse-like excitation to the vocal tract system. Inverse filtering of a speech signal can remove the vocal tract contribution from the speech signal, and this signal is known as a Linear prediction (LP) residual. In general, the LP residual signal is treated as an approximation of the excitation source signal [13]. Inverse LP analysis removes lower order relations present among the speech samples and retain the higher order relations. The relations present among the adjacent speech samples are treated as lower order relations and those among distant samples are treated as higher order relations. Due to presence of higher order relations, the LP residual signal appears like a random signal and does not contain information beyond the fundamental frequency of speech. The features derived from the LP residual may contain useful information, and it can be used for developing different speech systems. These source features contain supplementary information with respect to the vocal tract features, since the vocal tract system features mainly represent lower order relations present among the speech samples. The features used to represent glottal activity, mainly the vibration of glottal folds, are known as source or excitation source features. These terms are interchangeably used in this book, to represent source information. The LP residual of the voiced speech segment is observed to be a sequence of impulses corresponding to instants of glottal closure, where the prediction error is maximum. In the case of unvoiced speech, the magnitude of LP residual samples is relatively higher compared to LP residual samples of voiced speech. We propose the source information for characterizing the speech emotions, as the higher order relations present in the LP residual may contain emotion-specific information along with other speech features.

1.3.3 System Features

The sequence of impulse-like excitation, caused due to vocal folds' vibration, acts as a stimulus to the vocal tract system. The vocal tract system can be considered as a cascade of cavities of varying cross sections. The sequences of shapes assumed by the vocal tract, while producing different sound units, are treated as the vocal tract system characteristics of the sound units. During speech production, a vocal tract acts as resonator and emphasizes certain frequency components depending on the shape of the oral cavity. Formants are the resonances of the vocal tract system at a given instance of time. These formants are represented by bandwidth and amplitude, and these parameters are unique to each of the sound units. The sequence of shapes of the vocal tract system also carries emotion-specific information, along with the information related to the sound unit. These features are clearly observed in the frequency domain. For frequency domain analysis, a speech signal is segmented into frames of size 20–30 ms, with a frame shift of 10 ms. The spectrum of the speech frame is obtained through Fourier transform. From the magnitude spectrum, the features like linear prediction cepstral coefficients (LPCCs), mel frequency cepstral coefficients (MFCC's), perceptual linear prediction coefficients (PLPCs) and their derivatives are computed to represent vocal tract characteristics [14, 15]. Features extracted from the vocal tract system are generally known as spectral, system or segmental level features. These terms are interchangeably used in this book, to represent vocal tract information. MFCCs, LPCCs, PLPCs are widely known spectral features used in the literature for various speech tasks. The information about the sequence of shapes of the vocal tract, responsible for producing various is captured through spectral features [14, 15]. The peaks in the spectra are known as formants. They indicate the resonance frequencies of the vocal tract system.

1.3.4 Prosodic Features

Speech features extracted from longer speech segments like syllables, words and sentences are known as prosodic features. They represent overall quality of the speech such as rhythm, stress, intonation, loudness, emotion and so on. It is difficult to derive suitable correlates to represent the above mentioned speech quality features. In the literature, pitch, duration, energy and their derivatives are widely used to represent prosodic features. These prosodic features are also known as supra-segmental or long-term features. These terms are interchangeably used in this book, to represent prosodic information. Human beings perceive emotions present in speech, by exploiting the prosodic features, and in this studies, these features are explored for classifying the emotions. From the above discussion, the importance to explore the excitation source, vocal tract system and prosodic features, to capture emotion-specific information, is observed. This book addresses an issue of *speech emotion recognition* by exploring the above mentioned emotion-specific features, for discriminating the emotions. Terms such as recognition performance, classification performance, and discrimination are used in this book, in the context of emotions, unless specifically mentioned.

1.4 Emotional Speech Databases

An important issue to be considered in evaluating emotional speech systems is the quality of the databases used to develop and assess the performance of the systems [5]. The objectives and methods of collecting speech corpora highly vary according to the motivation behind the development of speech systems. Speech corpora used for developing emotional speech systems can be divided into 3 types. The important properties of these databases are briefly mentioned in Table 1.1.

1. Actor (Simulated) based emotional speech database.
2. Elicited (Induced) emotional speech database.
3. Natural emotional speech database.

Simulated emotional speech corpora are collected from reasonably experienced and trained theatre or radio artists. Artists are asked to express linguistically neutral sentences in different emotions. Recording is done in different sessions to consider the variations in the degree of expressiveness and physical speech production mechanism of human beings. This is one of the easier and reliable methods of collecting expressive speech databases containing a wide range of emotions. More than 60 % of the databases collected for expressive speech research are of this kind. The emotions collected through simulated means are fully developed in nature, which are typically intense, and incorporate most of the aspects considered relevant for the emotion [21]. These are also known as *full blown* emotions. Generally, it is found that acted/simulated emotions tend to be more expressed than real ones [5, 22].

Elicited speech corpora are collected by simulating the artificial emotional situation, without the knowledge of the speaker. Speakers are made to involve themselves in emotional conversation with the anchor, where different contextual situations are created by the anchor through conversation to elicit different emotions from the subject, without his/her knowledge. These databases may be more natural than their simulated counterparts, but subjects may not be properly expressive if they know

Table 1.1 Different types of databases used in speech emotion recognition

Type of database	Advantages	Disadvantages
Actor(Simulated) Eg: LDC speech corpus [16], Emo-DB [17], IITKGP-SESC [18]	• Most commonly used • Standardized • Results can be compared easily • Complete range of emotions is available • Wide variety of databases in most of the languages is available	• Acted speech tells how emotions should be portrayed rather than how they are portrayed • Contextual, individualistic and purpose dependent information is absent • Episodic in nature, which is not true in real world situation • Often it is read speech, not spoken
Elicited(Induced) Eg: Wizard of Oz databases, ORESTEIA [19]	• Nearer to the natural databases • Contextual information is present, but it is artificial	• All emotions may not be available • If the speakers know that they are being recoded, the quality will be artificial
Natural Eg: Call center conversations [20], Cockpit recordings	• They are completely naturally expressed • Mostly useful for real world	• All emotions may not be available • Copyright and privacy issues • Overlapping of utterances • Presence of background noise • Contains multiple and concurrent emotions • Pervasive in nature. So, difficult to model

that they are being recorded. Sometimes these databases are recorded by asking the subjects to involve themselves in verbal interaction with a computer whose speech responses are in turn controlled by the human being without the knowledge of the subject [23].

Unlike full blown emotions, natural emotions are mildly expressed. Sometimes, It may be difficult to clearly recognize these emotions. They are also known as *underlying emotions*. Naturally available real world data may be recorded from call center conversations, cockpit recordings during abnormal conditions, a dialogue between patient and a doctor and so on. However, it is difficult to find a wide emotion base in this category. Annotation of these emotions is also highly subjective (expert opinion based) and categorization is always debatable. There are also some legal issues, such as privacy and copyright, while using natural speech databases [5, 23]. Table 1.1 briefly explains the advantages and drawbacks of the three types of emotional speech databases.

1.5 Applications of Speech Emotion Recognition

Speech emotion recognition has several applications in day-to-day life. It is particularly useful for enhancing the naturalness in speech-based human machine interaction [18, 24, 25]. Emotion recognition systems may be used in an on-board car driving system, where information about the mental state of a driver may be used to keep him alert during driving. This helps to avoid some accidents, caused by a stressed mental state of the driver [24]. Call center conversation analysis may be helpful behavioral study of call attendants with the customers, and helps to improve the quality of service of a call attendant [20, 26]. Interactive movie [26], story telling [27] and E-tutoring [1] applications would be more practical, if they can adapt themselves to listeners' or students' emotional states. The automatic way to analyze the emotions in speech is useful for indexing and retrieving the audio/video files based on emotions [28]. Medical doctors may use the emotional contents of the patient's speech as a diagnosing tool for various disorders [29]. Emotion analysis of telephone conversation between criminals would help crime investigation departments. Conversation with robotic pets and humanoid partners would be more realistic and enjoyable, if they are able to express and understand emotions like humans [30]. It may also be useful in automatic speech-to-speech translation systems, where the speech in language x is translated into other language y by the machine. Here, both emotion recognition and synthesis are used. The emotions present in the source speech are to be recognized, and the same emotions are to be synthesized in the target speech, as the translated speech is expected to represent the emotional state of the original speaker [5]. In aircraft cockpits, speech recognition systems trained to recognize stressed-speech are used for better performance [31]. Call analysis in emergency services like ambulance and fire brigade may help to evaluate the genuineness of the requests.

1.6 Issues in Speech Emotion Recognition

Some of the important research issues in speech emotion recognition are discussed below in brief.

- The word *emotion* is inherently uncertain and subjective. The term *emotion* has been used with different contextual meanings by different people. It is difficult to define *emotion* objectively, as it is an individual mental state that arises spontaneously rather than through conscious effort. Therefore, there is no common objective definition and agreement on the term *emotion*. This is the fundamental hurdle to proceed with the research [32].

- There are no standard speech corpora for comparing the performance of research approaches used to recognize emotions. Most of the systems processing emotional speech are developed using full blown emotions, but real life emotions are pervasive and underlying in nature. Some databases are recorded using experienced artists, whereas some others are recorded using semi experienced or inexperienced subjects. The research on emotion recognition is limited to 5–6 emotions, as most of the databases do not provide a wide variety of emotions [16].
- The emotion recognition systems developed using various features may be influenced by the speaker and language specific information. Ideally, speech emotion recognition systems should be speaker and language independent [33].
- An important issue in the development of a speech emotion recognition systems is identification of suitable features that efficiently characterize different emotions [5]. Along with the features, suitable models are to be identified to capture emotion-specific information from the extracted speech features.
- Speech emotion recognition systems should be robust enough to process real-life and noisy speech to identify the emotions.

1.7 Objectives and Scope of the Book

The objective of this book is to explore various emotion-specific features from spectral and prosodic aspects of speech for developing a robust emotion recognition system. In the literature, most of the speech tasks are performed by processing entire speech signal, frame by frame for extracting the spectral features. However critical analysis of speech signals, motivated us to use spectral features only from sub-syllabic regions such as: vowels and consonants to develop emotion recognition systems, ensuring no serious decrease in the recognition performance. Therefore, in this work, we have investigated vowel, consonant and consonant to vowel transition regions separately for discriminating the emotions. Frame level processing of speech provides average spectral characteristics correspond to pitch cycles present in that frame. However, it is known that even successive pitch cycles in a speech signal vary with respect to time. For capturing these finer spectral variations present in adjacent pitch cycles of speech signal, we have proposed pitch synchronous spectral features for discriminating the emotions. In addition to the proposed spectral features, we have also explored global and local prosodic features at syllable and word levels for discriminating the emotions. The importance of syllable and word positions in discriminating the emotions is also investigated in this work. After exploring the source, system, and prosodic features independently for emotion recognition, various fusion approaches are explored for combining the evidence to enhance the recognition performance. A multi-stage classification approach has been explored for further improvement in the emotion recognition performance. In this work, we have also explored speaking rate characteristics for discriminating emotions. The proposed features are evaluated on both simulated and real life emotion speech databases.

1.8 Main Highlights of Research Investigations

- Proposing sub-syllabic regions such as vowel, consonant and consonant-vowel transition regions for extracting spectral features.
- Proposing pitch synchronous spectral features for discriminating the emotions.
- Proposing global and local prosodic features at word and syllable levels for emotion recognition.
- Investigating the role of syllable and word positions in discriminating the emotions.
- Studying different methods for combining the speech features to improve emotion recognition performance.
- Proposing a two-stage emotion recognition system for improving the recognition performance.
- Effectiveness of the proposed features are demonstrated on both simulated and real life emotion speech databases.

1.9 Brief Overview of Contributions in this Book

The major contributions of the book include exploring the emotion-specific features from the vocal tract system and prosodic components of speech. The details are discussed in the following sub-sections.

1.9.1 Emotion Recognition using Spectral Features Extracted from Sub-syllabic Regions and Pitch Synchronous Analysis

A conventional block processing approach is widely used in the literature to extract different system features. In this book, we proposed the vocal tract system features extracted from sub-syllabic segments like consonants, vowels and consonant-vowel (CV) transition regions, to recognize emotions from speech signals. From the analysis, it is observed that system features derived from consonant-vowel transition regions have achieved the emotion recognition performance close to the system features derived from the whole speech signal. Pitch synchronous analysis of speech signals is also proposed to capture fine and gradual variations of vocal tract characteristics, to discriminate emotions. Emotion recognition performance using pitch synchronously extracted system features achieved better results than the conventional block processing approach. Formant features in combination with other system features shown the improvement in emotion recognition performance.

1.9.2 Emotion Recognition using Global and Local Prosodic Features Extracted from Words and Syllables

Most of the existing emotion recognition studies have focused on global prosodic features. In this book, local (dynamic) prosodic features have been proposed for emotion recognition. The combination of global and local prosodic features is also used for discriminating the emotions. The contribution of different speech segments like syllables, words and sentences and their positions (initial, middle, and final) are also studied in detail, in the context of emotion recognition using prosodic features.

1.9.3 Emotion Recognition using Combination of Features

In this book, we have explored emotion-specific spectral and prosodic features for discriminating the emotions. The proposed features may provide some complementary or supplementary information with respect to each other. Therefore, for exploiting the complementary emotion-specific information provided by individual features, appropriate fusion techniques are proposed in this book for combining the evidences provided by various features.

1.9.4 Emotion Recognition on Real Life Emotional Speech Database

The proposed spectral and prosodic features are evaluated on real life emotional speech database. In this work, a Hindi movie database is used to represent real life emotions. The recognition performance using proposed features seems to be better compared to conventional spectral and prosodic features. A Two-stage emotion recognition approach based on speaking rate has been proposed for improving the emotion recognition performance further.

1.10 Organization of the Book

The evolution of ideas presented in this book is given at the end of the chapter in Table 1.2. The chapter-wise organization of the book is given below.

- **Chapter 1**: **The Introduction** discusses about the characterization of emotions by machines and human beings. The philosophy of emotions with respect to the psychology and speech production aspects is briefly discussed. Different types of emotional speech corpora, contemporary issues in speech emotion recognition, applications, objectives and scope of this book are briefly discussed. Evolution of the ideas of this book is given at the end of the chapter.

Table 1.2 Evolution of ideas presented in this book

Evolution of ideas presented in this book

- Most of the present speech systems are efficient in processing neutral speech
- Incorporation of components of *emotion processing* into today's speech systems makes them natural, robust and pragmatic
- Modeling *emotions* requires a suitable emotional speech corpus
  − Simulated emotional speech corpus in Telugu is collected using radio artists in 8 emotions
- Emotion-specific features need to be identified to develop speech emotion recognition models
- Different speech features are explored toward emotion recognition
  − Vocal tract system features
  − Prosodic features
- While extracting spectral features, generally a block processing approach is used, where the entire speech signal is processed frame by frame. It is our intuition that it may not be required to process the entire speech signal for feature extraction as:
  − Steady vowel region contains redundant spectral information
  − Spectral variations are observed in CV transition regions
    These spectral variations may be discriminative in the case of different emotions
  − Consonant, vowel, and CV-transition regions are considered as the units of feature extraction, for emotion recognition
- In the case of conventional block processing, the speech signal is assumed to be stationary, within a frame of 20 ms (containing 3–4 pitch cycles). However expression of emotions is a gradually evolving phenomenon, hence, spectral features derived from the individual pitch cycles may contain finer variations, which can be explored for emotion recognition
  − Spectral features are pitch synchronously extracted and used for emotion recognition
- High amplitude regions of spectrum (formants) are proved to be robust in the case of noisy and speaker independent speech applications
  − Formant features are used in combination with cepstral features to develop emotion recognition systems
- Gaussian mixture models (GMMs) are well known to capture distributions of input data points Therefore, GMMs are employed to develop emotion recognition models using spectral features
- Prosodic information is treated to be the major contributor toward characterization of emotions
  Static (global) prosodic features are thoroughly explored in the literature for emotion recognition
  Dynamic nature of prosodic contours is observed to be more emotion-specific
  Different parts of the utterance may not contain emotion-specific information uniformly, as duration, pitch, and energy profiles vary with respect to the context
  − Dynamics of prosodic features is derived from prosodic contours
  − Prosodic features are extracted from sentences, words and syllables and used for emotion recognition
- Support vector machines (SVMs) are well known for capturing the discriminative information present among the feature vectors. Performance of SVMs is critical on the number of discriminative feature vectors (known as support vectors) rather than the total number of feature vectors. Therefore, SVMs are employed to develop emotion recognition models using prosodic features. There are less number of feature vectors in the case of prosodic analysis of emotions

Table 1.2 Continued

• The combination of evidences from different features is proved to perform better for many speech tasks
  This may be due to supplementary or complementary evidences provided by different features. Hence, in this work the following combination of features may be explored to study emotion recognition performance
  – Excitation and spectral features
  – Spectral and prosodic features
  – Excitation and prosodic features
  – Excitation, spectral and prosodic features
• Multilevel classification systems provide better classification over single level classification
  – Two level emotion classification system is proposed
  – At the first level all the emotions are divided into few broad groups, where similar (confusable) emotions are placed in different groups
  – At the second level emotions in broad groups are further classified
• The ultimate goal of any speech emotion recognition system is to process real life emotions
  – Combination of different features may be used for real-life emotion recognition
  – Hindi movie clips may be used to represent real life emotions

• **Chapter 2: Robust Emotion Recognition using Pitch Synchronous and Sub-syllabic Spectral Features** provides the details about the extraction of spectral features from sub-syllabic regions such as consonant, vowel, and consonant-vowel (CV) transition regions. Extraction of spectral features from pitch synchronous analysis is also explained. Development of emotion recognition systems using Gaussian mixture models is discussed.

• **Chapter 3: Robust Emotion Recognition using Sentence, Word and Syllable Level Prosodic Features** discusses in detail about the use of global and local prosodic features for developing emotion recognition systems. Global (static) and local (dynamic) prosodic features extracted from sentences, words, and syllables are proposed for classifying the speech emotions. The contribution of prosodic features from different speech regions (initial, middle, and final) is also analyzed using local and global features. For capturing emotion-specific prosody from the proposed features, support vector machine models are used.

• **Chapter 4: Robust Emotion Recognition using a Combination of Excitation Source, Spectral and Prosodic Features** discusses the combination of complementary and supplementary evidences provided by the source, system, and prosodic features for improving the emotion recognition performance. This chapter provides emotion recognition performance studies for various combinations of features. Here, evidences from various features are combined using optimal linear weighted combination scheme.

• **Chapter 5: Robust Emotion Recognition using Speaking Rate Features** discusses the development of two stage emotion recognition system based on speaking rate. In this case initially emotions are classified into broad categories like active (fast), passive (slow), and neutral (medium) based on speaking rate. Later, finer classification into individual emotions is performed.

- **Chapter 6: Emotion Recognition on Real Life Emotions** discusses the effectiveness of the proposed spectral and prosodic features in recognizing real life emotions. In this work, audio clips of emotional scenes in Hindi movies represent real life emotions.
- **Chapter 7: Summary and Conclusions** summarizes the contributions of this book along with some important conclusions. This chapter also provides the scope for the extensions to the present work and future directions for improving the performance of emotion recognition models.

References

1. D. Ververidis, C. Kotropoulos, A state of the art review on emotional speech databases, in *Eleventh Australasian International Conference on Speech Science and Technology*, Auckland, New Zealand, Dec 2006
2. S.G. Koolagudi, N. Kumar, K.S. Rao, Speech emotion recognition using segmental level prosodic analysis, in *International Conference on Devices and Communication*, Birla Institute of Technology, Mesra, India, Feb 2011. (IEEE Press, Washington DC, 2011)
3. M. Schubiger, *English Intonation: Its Form and Function* (Niemeyer, Tubingen, 1958)
4. J. Connor, G. Arnold, *Intonation of Colloquial English*, 2nd edn, (Longman, London, 1973)
5. M.E. Ayadi, M.S. Kamel, F. Karray, Survey on speech emotion recognition: features, classification schemes, and databases. Pattern Recognit. **44**, 572–587 (2011)
6. P. Ekman, Basic emotions, in *Handbook of Cognition and Emotion* (Wiley, Sussex, 1999)
7. R. Cowie, E. Douglas-Cowie, N. Tsapatsoulis, S. Kollias, W. Fellenz, J. Taylor, Emotion recognition in human–computer interaction. IEEE Signal Process. Mag. **18**, 32–80 (2001)
8. J. William, What is an emotion? Mind **9**, 188–205 (1884)
9. A.D. Craig, Interoception and emotion: a neuroanatomical perspective, in *Handbook of Emotion* (The Guildford Press, New York, 2009). ISBN: 978-1-59385-650-2
10. C.E. Williams, K.N. Stevens, Vocal correlates of emotional states, in *The evaluation of speech in psychiatry* (Grune and Stratton Inc., New York, 1981), pp. 189–220
11. J. Cahn, The generation of affect in synthesized speech. J. Am. Voice Input/Output Soc. **8**, 1–19 (1990)
12. G.M. David, Theories of emotion, in *Psychology*, vol. 7 (Worth publishers, New York, 2004)
13. J. Makhoul, Linear prediction: a tutorial review. Proc. IEEE **63**(4), 561–580 (1975)
14. L.R. Rabiner, B.H. Juang, *Fundamentals of Speech Recognition* (Prentice-Hall, Englewood Cliffs, 1993)
15. J. Benesty, M.M. Sondhi, Y. Huang (eds.), *Springer Handbook on Speech Processing* (Springer, Berlin, 2008)
16. D. Ververidis, C. Kotropoulos, Emotional speech recognition: resources, features, and methods. Speech Commun **48**, 1162–1181 (2006)
17. F. Burkhardt, A. Paeschke, M. Rolfes, M. Sendlmeier, B. Weiss, A database of german emotional speech, in *Interspeech*, 2005
18. S.G. Koolagudi, S. Maity, V.A. Kumar, S. Chakrabarti, K.S. Rao, in *IITKGP-SESC: Speech Database for Emotion Analysis*, Communications in Computer and Information Science, JIIT University, Noida, India, 17–19 Aug 2009, Springer. ISSN: 1865–0929
19. E. McMahon, R. Cowie, S. Kasderidis, J. Taylor, S. Kollias, What chance that a dc could recognize hazardous mental states from sensor inputs?, in *Tales of the Disappearing Computer*, Santorini, Greece, 2003

20. C.M. Lee, S.S. Narayanan, Toward detecting emotions in spoken dialogs. IEEE Trans. Speech Audio Process. **13**, 293–303 (2005)
21. M. Schroder, R. Cowie, E. Douglas-Cowie, M. Westerdijk, S. Gielen, Acoustic correlates of emotion dimensions in view of speech synthesis, in *7th European Conference on Speech Communication and Technology*, EUROSPEECH 2001 Scandinavia, 2nd INTERSPEECH Event, Aalborg, Denmark, 3–7 Sept 2001
22. C. Williams, K. Stevens, Emotions and speech: some acoustical correlates. J. Acoust. Soc. Am. **52**(4 pt 2), 1238–1250 (1972)
23. A. Batliner, J. Buckow, H. Niemann, E. Noth, V. Warnke, *Verbmobile Foundations of Speech to Speech Translation* (Springer, Berlin, 2000). ISBN: 3540677836, 9783540677833
24. B. Schuller, G. Rigoll, M. Lang, Speech emotion recognition combining acoustic features and linguistic information in a hybrid support vector machine-belief network architecture, in *Proceedings of the IEEE International Conference on Acoustics, Speech, and Signal Processing, (ICASSP '04)*, May 17–21. (IEEE Press, 2004), pp. I-577–580. ISBN: 0-7803-8484-9
25. F. Dellert, T. Polzin, A. Waibel, Recognizing emotion in speech, in *4th International Conference on Spoken Language Processing*, pp. 1970–1973, Philadelphia, 3–6 Oct 1996
26. R. Nakatsu, J. Nicholson, N. Tosa, Emotion recognition and its application to computer agents with spontaneous interactive capabilities. Knowl.-Based Syst. **13**, 497–504 (2000)
27. F. Charles, D. Pizzi, M. Cavazza, T. Vogt, E. Andr, Emoemma: emotional speech input for interactive storytelling, in *8th International Conference on Autonomous Agents and Multiagent Systems (AAMAS 2009)*, pp. 1381–1382, Budapest, Hungary, 10–15 May 2009
28. T.V. Sagar, Characterisation and synthesis of emotionsin speech using prosodic features, Master's thesis, Department of Electronics and Communications Engineering, Indian Institute of Technology, Guwahati, May 2007
29. D.J. France, R.G. Shiavi, S. Silverman, M. Silverman, M. Wilkes, Acoustical properties of speech as indicators of depression and suicidal risk. IEEE Trans. Biomed. Eng. **47**(7), 829–837 (2000)
30. P.-Y. Oudeyer, The production and recognition of emotions in speech: features and algorithms. Int. J. Human Comput. Stud. **59**, 157–183 (2003)
31. J. Hansen, D. Cairns, Icarus: source generator based real-time recognition of speech in noisy stressful and lombard effect environments. Speech Commun. **16**(4), 391–422 (1995)
32. M. Schroder, R. Cowie, Issues in emotion-oriented computing toward a shared understanding, in *Workshop on Emotion and Computing (HUMAINE)*, 2006
33. S.G. Koolagudi, K.S. Rao, Real life emotion classification using vop and pitch based spectral features, in *INDICON-2010*, Jadavpur University, Kolkata, India, Dec 2010

Chapter 2
Robust Emotion Recognition using Pitch Synchronous and Sub-syllabic Spectral Features

Abstract This chapter discusses the use of vocal tract information for recognizing the emotions. Linear prediction cepstral coefficients (LPCC) and mel frequency cepstral coefficients (MFCC) are used as the correlates of vocal tract information. In addition to LPCCs and MFCCs, formant related features are also explored in this work for recognizing emotions from speech. Extraction of the above mentioned spectral features is discussed in brief. Further extraction of these features from sub-syllabic regions such as consonants, vowels and consonant-vowel transition regions is discussed. Extraction of spectral features from pitch synchronous analysis is also discussed. Basic philosophy and use of Gaussian mixture models is discussed in this chapter for classifying the emotions. The emotion recognition performance obtained from different vocal tract features is compared. Proposed spectral features are evaluated on Indian and Berlin emotion databases. Performance of Gaussian mixture models in classifying the emotional utterances using vocal tract features is compared with neural network models.

2.1 Introduction

In Chap. 1, we have introduced the topic of emotion recognition from speech using emotion specific speech features. In this chapter, we intend to discuss the use of vocal tract system features for speech emotion recognition. Features extracted from vocal tract system are generally known as spectral or system features. They are also sometimes referred to as segmental features, as they are normally extracted from the speech segments of 20–30 ms. MFCCs (Mel frequency cepstral coefficients), LPCCs (Linear prediction cepstral coefficients), perceptual linear prediction coefficients (PLPCs) are widely known spectral features used in the literature [1].

Generally, spectral features are found to be robust for various speech tasks. This may be due to the accurate representation of vocal tract system characteristics by spectral features. Figure 2.1 shows the unique spectral characteristics for 8 emotions

K. S. Rao and S. G. Koolagudi, *Robust Emotion Recognition using Spectral and Prosodic Features*, SpringerBriefs in Speech Technology,
DOI: 10.1007/978-1-4614-6360-3_2, © The Author(s) 2013

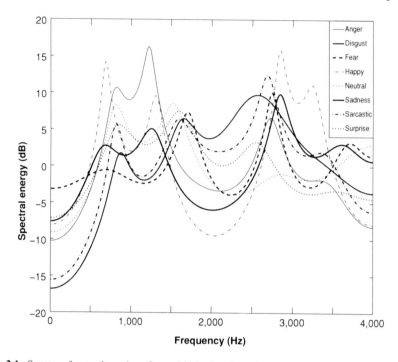

Fig. 2.1 Spectra of a steady region of vowel /A/, taken from the utterance *anni dAnamulalo vidyA dAnamu minnA*, in 8 emotions

of IITKGP-SESC. The spectra shown in Fig. 2.1 represent the steady region of a vowel /A/ from Telugu utterance *anni dAnamulalo vidyA dAnamu minnA*, expressed in eight different emotions. It is observed from the figure that the sharpness of the formant peaks, positions of the formants, formant bandwidths, and spectral tilt have distinctive properties for different emotions. In the literature, spectral features are used for modeling the pitch pattern of the speaker [2]. Variation of the pitch is an important correlate of emotions in speech. It is also known that variation in the pitch leads to changes in other prosodic parameters like duration and energy. Intuitively it may be harder to correlate spectral features with the temporal dynamics in prosody, related to emotional states. However they provide a detailed description of varying speech signal, including variations in prosody [3]. Our intuition here is that spectral features capture not only the information about what is being said (text) but also how it is being said (emotion). Therefore, spectral features are used in this study to recognize speech emotions.

The majority of the works on speech emotion recognition are carried out using spectral features derived from the whole speech signal through a conventional block processing approach. In this chapter, we have proposed the spectral features from sub-syllabic regions of the speech for recognizing the emotions. Generally, a syllable consists of a vowel optionally preceded or/and followed by consonants (this of course, varies across languages). Vowel is treated as the nucleus of the syllable. In this context, we have used vowel onset points (VOPs) to identify consonant, vowel and consonant-to-vowel transition regions from a CV unit. In this work, these three regions are referred to as sub-syllabic regions. Features extracted from these three regions are used to recognize the emotions from a speech utterance. Most of the speech signal contains vowel regions. Steady portions of the vowel, mostly contain redundant spectral information. In this context, there is no need to process entire speech signal for feature extraction, to perform speech emotion recognition. This chapter will show that the approach of processing only useful portions from speech signal for feature extraction increases the desired performance of the system and reduces the time and computational complexities.

In case of a block processing approach, the speech signal is assumed to be stationary within a considered block of 20 ms, which includes multiple pitch cycles. In reality, the speech signal is not stationary, as it is the outcome of a time varying vocal tract system excited by the time varying excitation source. Even two consecutive cycles of a speech signal are slightly different. Generally, manifestation of emotions through speech is observed to be a gradual process, leading to emotion-specific modulation, during production of speech. Hence, instead of using the average spectrum over multiple pitch cycles, the spectrum derived from each pitch cycle is considered for feature extraction. Variations in instantaneous pitch information may also be captured using pitch synchronous spectral features. Therefore analyzing spectral features of each cycle (pitch period) of a speech signal may provide more emotion distinctive information than the emotion-specific information obtained from a block processing approach. This approach of analyzing a speech signal considering one pitch cycle each time is known as pitch synchronous analysis. Specifically no evidence is found in the literature, for pitch synchronously analyzing the speech signal to recognize emotions. In this chapter, pitch synchronously extracted spectral features are explored for capturing emotion-specific information. The proposed approach is verified using two databases: IITKGP-SESC and Emo-DB. Details of these databases are explained in Sect. 2.2.

The remaining part of the chapter is organized as follows, Sect. 2.2 briefs about the two databases used in this work. Extraction of different types of spectral features mentioned above is discussed in Sect. 2.3. Section 2.4 discusses the methodology of Gaussian mixture models used to develop emotion recognition models. Emotion recognition results are discussed in Sect. 2.5. Chapter concludes with a summary.

2.2 Emotional Speech Corpora

In this work, the Indian Institute of Technology Kharagpur- Simulated Emotional Speech Corpus (IITKGP-SESC) and Berlin emotional speech database (Emo-DB) are used to analyze the performance of proposed spectral features for Emotion Recognition (ER).

2.2.1 Indian Institute of Technology Kharagpur-Simulated Emotional Speech Corpus: IITKGP-SESC

This database is particularly designed and developed at the Indian Institute of Technology, Kharagpur, to support the study on speech emotion recognition. The proposed speech database is the first one developed in an Indian language (Telugu), for analyzing the common emotions present in day-to-day conversations. This corpus is sufficiently large to analyze the emotions in view of speaker, gender, text and session variability.

The corpus is recorded using 10 (5 male and 5 female) professional artists from All India Radio (AIR) Vijayawada, India. The artists are sufficiently experienced in expressing the desired emotions from the neutral sentences. All the artists are in the age group of 25–40 years, and have the professional experience of 8–12 years. For analyzing the emotions we have considered 15 semantically neutral Telugu sentences. Each of the artists has to speak the 15 sentences in 8 given emotions in one session. The number of sessions considered for preparing the database is 10. The total number of utterances in the database is 12000 (15 *sentences* × 8 *emotions* × 10 *artists* × 10 *sessions*). Each emotion has 1500 utterances. The numbers of words and syllables in the sentences are varying from 3–6 and 11–18 respectively. The total duration of the database is around 7 h. The eight emotions considered for collecting the proposed speech corpus are: Anger, Disgust, Fear, Happiness, Neutral, Sadness, Sarcasm and Surprise. The speech samples are recorded using SHURE dynamic cardioid microphone C660N. The distance between the microphone and the speaker is maintained approximately around 3–4 inches. The speech signal is sampled at 16 kHz, and each sample is represented as a 16-bit number. The sessions are recorded on alternate days to capture the inevitable variability in the human vocal tract system. In each session, all the artists have given the recordings of 15 sentences in 8 emotions. The recording is done in such a way that each artist has to speak all the sentences at a stretch in a particular emotion. This provides the coherence among the sentences for each emotion category. The entire speech database is recorded using a single microphone and at the same location. The recording was done in a quiet room, without any obstacles in the recording path [4].

The quality of the database is also evaluated using subjective listening tests. Here, the quality represents how well the artists simulated the emotions from the neutral text. The subjects are used to assess the naturalness of the emotions embedded in speech utterances. This evaluation is carried out by 25 post graduation and research students of the Indian Institute of Technology, Kharagpur. This subjective listening test is useful for the comparative analysis of emotions in a human versus machine perspective. In this study, 40 sentences (5 sentences from each emotion) randomly selected from male and female speakers are considered for evaluation. Before taking the test, the subjects are given the pilot training by playing 8 sentences (a sentence from each emotion) from each artist's speech data, for understanding (familiarizing) the characteristics of emotion expression. Forty sentences used in this evaluation are randomly ordered, and played to the listeners. For each sentence, the listener has to

Table 2.1 Emotion classification performance based on subjective evaluation

	Male Artist								Female Artist							
	A	D	F	H	N	Sa	S	Sur	A	D	F	H	N	Sa	S	Sur
Anger	73	17	2	3	4	0	0	1	69	19	3	2	5	0	0	2
Disgust	28	56	7	0	4	0	5	0	40	44	5	0	3	2	3	3
Fear	7	6	49	0	8	19	1	10	6	8	37	2	7	25	1	14
Happiness	0	2	6	62	8	9	5	6	0	4	4	66	10	7	3	6
Neutral	0	5	0	6	86	2	0	1	1	8	1	6	83	0	1	0
Sadness	0	3	16	3	13	61	4	0	4	2	25	1	12	52	3	1
Sarcasm	6	5	0	0	0	4	85	0	4	5	0	0	0	3	88	0
Surprise	0	7	5	16	5	5	7	55	6	6	3	17	3	16	1	48

A Anger, *D* Disgust, *F* Fear, *H* Happiness, *N* Neutral, *Sa* Sadness, *S* Sarcasm, *Sur* Surprise

mark the emotion category from the set of 8 given emotions. The overall emotion classification performance for male and female speech data is given in Table 2.1. The observation shows that the average emotion recognition rates of male and female speech utterances are 61 and 66 % respectively.

In this book, emotion recognition studies have been carried out in three ways, accordingly three data sets are derived from IITKGP-SESC. They are (1) Set1: Session independent speech emotion recognition, (2) Set2: Text independent speech emotion recognition, and (3) Set3: Speaker and text independent speech emotion recognition. Set1 is used to analyze the emotion recognition in view of session variability. Here 8 sessions of all speakers' speech data is used for training the emotion models, and the remaining 2 sessions of all speakers' speech data is used for testing. Set2 is used to study emotion recognition in view of text independent speech data. Here, 8 sessions of all speakers' speech data containing the first 10 text prompts are used for training, and while testing the remaining 5 text prompts of the last 2 sessions from all the speakers' speech data are used. Set3 is used to analyze the emotion recognition with respect to speaker and text independent speech data. Here, training is performed with 8 speakers' (4 males and 4 females) speech data, from all 10 sessions. Testing is performed with the remaining 2 speakers' (one male and one female) speech data. To realize the text independent data, during training the speech utterances corresponding to the first 10 text prompts of the database are used, and the remaining 5 text prompts are used while testing. Set3 is designed to have text and speaker independent properties and is more generalized than the other 2 sets. Therefore the majority of the results discussed in this book are derived using the Set3 dataset. Brief results of Set1 and Set2 are discussed at the end of the respective chapters. Table 2.2 shows the details of the three datasets used in this work.

2.2.2 Berlin Emotional Speech Database: Emo-DB

F. Burkhardt et al. have collected the actor-based simulated-emotion Berlin database in the German language [5]. Ten (5 male + 5 female) actors have contributed in preparing the database. The emotions recorded in the database are anger, boredom,

Table 2.2 Details of the datasets derived from IITKGP-SESC, for various studies on speech emotion recognition

Data set	Purpose and description	Training data	Testing data
Set1	Session independent emotion recognition	The utterances of all 15 text prompts, recorded from 10 speakers are used in training. Out of 10 sessions, 8 (1–8) sessions of each speaker are used in training.	The utterances of all 15 text prompts, recorded from 10 speakers are used in testing. Out of 10 sessions, 2 (9 and 10) sessions of each speaker are used in testing.
Set2	Session and text independent emotion recognition	Out of 15 text prompts, the utterances of 10 (1–10) prompts, recorded from 10 speakers are used in training. Out of 10 sessions, 8 (1–8) sessions of each speaker are used in training.	Out of 15 text prompts, the utterances of 5 (11–15) prompts, recorded from 10 speakers are used in testing. Out of 10 sessions, 2 (9 and 10) sessions of each speaker are used in testing.
Set3	Session, text, and speaker independent emotion recognition	Out of 15 text prompts, the utterances of 10 (1–10) prompts, recorded from 8 (4 males and 4 females) speakers are used in training. All 10 sessions of each speaker are used in training.	Out of 15 text prompts, the utterances of 5 (11–15) prompts, recorded from 2 (a male and a female) speakers are used in testing. All 10 sessions of each speaker are used in testing.

disgust, fear, happiness, neutral and sadness. Ten linguistically neutral German sentences are chosen for database construction. The database is recorded using the Sennheiser MKH 40 P48 microphone, with the sampling frequency of 16 kHz. Samples are stored as 16 bit numbers. Eight hundred and forty (840) utterances of Emo-DB are used in this work.

In the case of the Berlin database, 8 speakers' speech data is used for training the models and the remaining 2 speakers' speech data is used for validating the trained models.

2.3 Feature Extraction

In this work, LPCCs, MFCCs and formant features are used for representing the spectral information. Sub-syllabic spectral features are derived from the speech segments of consonants, vowels, and consonant to vowel transitions. Pitch synchronous spectral features are derived from each pitch cycle of the speech signal. Extraction of different spectral features, mentioned above is discussed in the following subsections.

2.3.1 Linear Prediction Cepstral Coefficients (LPCCs)

The cepstral coefficients derived from either linear prediction (LP) analysis or a filter bank approach are almost treated as standard front end features. Speech systems developed based on these features have achieved a very high level of accuracy, for speech recorded in a clean environment. Basically, spectral features represent phonetic information, as they are derived directly from spectra. The features extracted from spectra, using the energy values of linearly arranged filter banks, equally emphasize the contribution of all frequency components of a speech signal. In this context, LPCCs are used to capture emotion-specific information manifested through vocal tract features. In this work, the 10th order LP analysis has been performed, on the speech signal, to obtain 13 LPCCs per speech frame of 20 ms using a frame shift of 10 ms. The human way of emotion recognition depends equally on two factors, namely: its expression by the speaker as well as its perception by a listener. The purpose of using LPCCs is to consider vocal tract characteristics of the speaker, while performing automatic emotion recognition [6].

Cepstrum may be obtained using linear prediction analysis of a speech signal. The basic idea behind linear predictive analysis is that the nth speech sample can be estimated by a linear combination of its previous p samples as shown in the following equation.

$$s(n) \approx a_1 s(n-1) + a_2 s(n-2) + a_3 s(n-3) + \cdots + a_p s(n-p)$$

where $a_1, a_2, a_3 \cdots$ are assumed to be constants over a speech analysis frame. These are known as predictor coefficients or linear predictive coefficients. These coefficients are used to predict the speech samples. The difference of actual and predicted speech samples is known as an error. It is given by

$$e(n) = s(n) - \hat{s}(n) = s(n) - \sum_{k=1}^{p} a_k s(n-k)$$

where $e(n)$ is the error in prediction, $s(n)$ is the original speech signal, $\hat{s}(n)$ is a predicted speech signal, a_ks are the predictor coefficients.

To compute a unique set of predictor coefficients, the sum of squared differences between the actual and predicted speech samples has been minimized (error minimization) as shown in the equation below

$$E_n = \sum_{m} \left[s_n(m) - \sum_{k=1}^{p} a_k s_n(m-k) \right]^2$$

where m is the number of samples in an analysis frame. To solve the above equation for LP coefficients, E_n has to be differentiated with respect to each a_k and the result is equated to zero as shown below

$$\frac{\partial E_n}{\partial a_k} = 0, \qquad \text{for } k = 1, 2, 3, \cdots, p$$

After finding the a_ks, one may find cepstral coefficients using the following recursion.

$$C_0 = \log_e p$$

$$C_m = a_m + \sum_{k=1}^{m-1} \frac{k}{m} C_k a_{m-k}, \qquad \text{for } 1 < m < p \text{ and}$$

$$C_m = \sum_{k=m-p}^{m-1} \frac{k}{m} C_k a_{m-k}, \qquad \text{for } m > p$$

2.3.2 Mel Frequency Cepstral Coefficients

A human auditory system is assumed to process a speech signal in a nonlinear fashion. It is well known that lower frequency components of a speech signal contain more phoneme specific information. Therefore a nonlinear mel scale filter bank has been used to emphasize lower frequency components over higher ones. In speech processing, the mel frequency cepstrum is a representation of the short term power spectrum of a speech frame using a linear cosine transform of the log power spectrum on a nonlinear mel frequency scale. Conversion from normal frequency f to mel frequency m is given by the equation

$$m = 2595 \log_{10} \left(\frac{f}{700} + 1 \right)$$

The steps used for obtaining mel frequency cepstral coefficients (MFCCs) from a speech signal are as follows:

1. Pre-emphasize the speech signal.
2. Divide the speech signal into a sequence of frames with a frame size of 20 ms and a shift of 5 ms. Apply the hamming window over each of the frames.
3. Compute the magnitude spectrum for each windowed frame by applying DFT.
4. Mel spectrum is computed by passing the DFT signal through a mel filter bank.
5. DCT is applied to the log mel frequency coefficients (log mel spectrum) to derive the desired MFCCs.

Twenty filter banks are used to compute 8, 13 and 21 MFCC features from a speech frame of 20 ms, with 5 ms overlap each time. The purpose of using MFCCs is to take the listener's non-linear auditory perceptual system into account, while performing automatic emotion recognition.

2.3.3 Formant Features

Cepstral coefficients are used as standard front end features for developing various speech systems, however, they perform poorly with noisy or real life speech. Therefore the supplementary features along with basic cepstral coefficients are essential to handle real life speech. The higher amplitude regions of a spectrum, such as formants, are relatively less affected by the noise. K. K. Paliwal et al. have extracted spectral sub-band centroids from high amplitude regions of the spectrum and used for noisy speech recognition [7]. With this viewpoint, formant parameters are used in this study as the supplementary features to cepstral features. Also note that the conventional cepstral features utilize only amplitude (energy) information from the speech power spectrum, whereas the proposed formant features utilize frequency information as well.

In general, formant tracks represent the sequences of vocal tract shapes, hence formant analysis using their strength, location and bandwidth may help to extract vocal tract related emotion specific information from a speech signal. Figure 2.2 shows different spectra for 8 emotions of IITKGP-SESC. The spectra are derived from the syllable *tha* from Telugu sentence *thallidhandrulanu gauravincha valenu*. In this case, the language, text, speaker and contextual information is maintained the same. This is speculative from the figure that the variation in the spectra is due to the emotions. Formant frequencies are very crucial in view of speech perception. Hence, a slight change in these parameters causes a perceptual difference, which may lead to manifestation of different emotions. It is evident from Fig. 2.2 that the position and strength of formants are clearly distinct for different emotions. Spectral peaks indicate the intensity of specific frequency components (or frequency band). Their distinctive nature for different emotions is the indication of presence of emotion specific information. The rate of decrease in spectrum amplitude, as a function of frequency is known as spectral roll-off or spectral tilt. This happens mainly because of decreasing strength of harmonics, as the frequency increases. A speaker can induce more strength into higher harmonics by consciously controlling the glottal vibration. Abrupt closing of the glottis increases the energy in the higher frequency components. This leads to the variation in spectral roll-off for different emotions. Figure 2.2 shows distinct spectral roll-offs for each of the emotions.

Though it is assumed that the bandwidth of a formant does not influence phonetic information [6], it represents some speaker specific information. Figure 2.2 depicts the variation in the formant bandwidths in case of different emotions. Even a slight variation in the bandwidth may be due to speaker induced emotion specific information as speaker, text, language and context related information do remain the same. Formant bandwidth is the frequency band measured at around 3 dB downward from the respective formant peak.

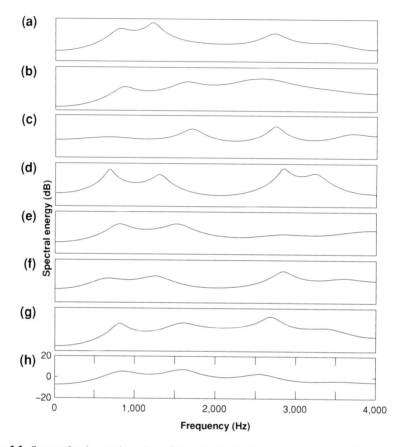

Fig. 2.2 Spectra for the steady region of the syllable /tha/ from the utterance *thallidhandrulanu gauravincha valenu*. **a** Anger, **b** Disgust, **c** Fear, **d** Happy, **e** Neutral, **f** Sadness, **g** Sarcastic, and **h** Surprise

2.3.4 Extraction of Sub-syllabic Spectral Features

In the context of Indian languages, a syllable can be viewed as the combination of consonants and a vowel, in the form $C^m V C^n$, where C is the consonant and V is the vowel such that, $m, n \geq 0$ and $m, n \leq 3$. In the syllable, the vowel is treated as the nucleus and consonants may or may not be present. In the context of Indian languages, most common syllable forms are CV, CCV, CCVC, and CVC. Among these forms, more than 90 % of the syllables are of the type CV. In a CV unit, the speech signal to the left of the VOP (before VOP) is a consonant region and to the right of it is the vowel region [8]. Vowel onset point as a junction point between the consonant and vowel of a CV unit. At this point the characteristics of the consonant segment are terminated and the characteristics of vowel are originated. Hence, it is important to extract features around this crucial point [9]. After determining the VOP, 40 ms to the left of the VOP

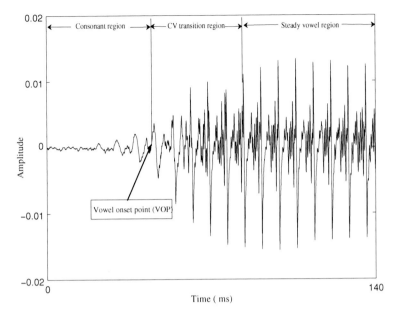

Fig. 2.3 Vowel onset point (consonant, vowel and CV transition regions are marked using vowel onset point) of a CV unit

is marked as the consonant region, and to the right of the VOP is the vowel region. In the vowel region, a small region of about 30 ms, following the VOP, is treated as transition region. After the transition region, 60 ms of speech signal is considered as the steady vowel region. Figure 2.3 shows the identified consonant, vowel, and CV transition regions from a typical CV unit. In this study, consonant, vowel and CV transition regions of a syllable are referred to as sub-syllabic speech regions. The features are extracted from vowel, consonant and transition regions, to analyze speech emotions. Processing complexity may be considerably reduced by avoiding processing of redundant information present in the speech regions such as the steady portion of the vowels. The purpose of this analysis is to study the contribution of different sub-syllabic regions toward emotion-specific information and also to avoid processing redundant speech regions from feature extraction. Therefore in this work, spectral (LPCCs, MFCCs and formant) features are extracted from consonant, vowel and CV transition regions of each syllable separately, using VOP locations as the anchor points.

VOP detection method employed in this work, uses the combination of measures from excitation source, spectral peaks, and modulation spectrum. This method is known as the combined method for the detection of VOP. Excitation source information is represented using the Hilbert envelope (HE) of the linear prediction (LP) residual. A sequence of the sum of the ten largest peaks of the spectra of speech frames represents the shape of the vocal tract. The slowly varying temporal envelope of the speech signal can be represented using a modulation spectrum. Each of these three features represents supplementary information about the VOP, and hence they are combined for the enhancement in the performance of VOP detection. VOP detection using the combined method is carried out with the following steps: (1) Derive the VOP evidence from the excitation source, spectral peaks, and modulation spectrum. Here, the evidence from excitation source information is obtained from the Hilbert envelope of the linear prediction residual signal. The evidence from the spectral peaks is obtained by summing the ten largest spectral peaks of each speech frame. The evidence due to the modulation spectrum is derived by passing the speech signal through a set of critical band pass filters, and summing the components corresponding to 4–16 Hz. (2) The above evidence is further enhanced by computing the slope with the help of a first order difference (FOD). (3) These enhanced parameters are convolved with the first order Gaussian difference (FOGD) operator for deriving the final VOP evidence. (4) Individual VOP parameters derived from the excitation source, spectral peaks and modulation spectrum are combined to provide the robust VOP evidence plot. (5) The positive peaks in the combined VOP evidence signal are hypothesized as the locations of VOPs [10]. Figure 2.4 shows the intermediate steps of VOP detection using the evidence from excitation source, spectral and modulation spectrum energy plots. More than 90 % of the automatically located VOPs are observed to be accurate with a maximum tolerance of 40 ms [10]. The sentence *Don't ask me to carry an oily rag like that*, chosen from TIMIT database, is used for illustrating the automatic detection of VOPs in Fig. 2.4. From the figure, it is observed that, the detected VOPs are close to the manually marked VOPs (See Figs. 2.4a, e).

The crucial part in CV units (syllable) is a region that represents transition from consonant to vowel. Figure 2.5 indicates the spectra obtained from the CV transition region of 40 ms length, for different emotions. The text, gender, speaker and contextual information is kept the same. From the figure, it may be observed that, except for the first formant, there is a clear distinction in the spectral characteristics for different emotions. For formant positions and energies are varying with respect to the emotions. Therefore, spectral characteristics derived from the CV transition regions of the speech signal are useful for discriminating the emotions. However, there is not much variation in the spectral characteristics of a speech signal in the steady portion of vowel. Consonant portion seem like noise containing frequency components with lower energy. The parameters extracted from these regions may not be much discriminative while capturing emotion-specific vocal tract characteristics. Figure 2.6 shows the spectra obtained from vowel, transition and consonant regions. It is clearly observed from the figure that the spectral behavior of a speech signal is mostly redundant in the case of vowels. Each frame in the transition region displays different spectral characteristics. The consonant region does not even show the formant structure, and energy of the spectral peaks is also very much less. To study this phenomenon, in this work, we have extracted LPCCs, MFCCs and formant features from 40 ms of each of CV transition and constant regions. Similarly, 60 ms of speech is considered from the steady vowel regions to extract the features.

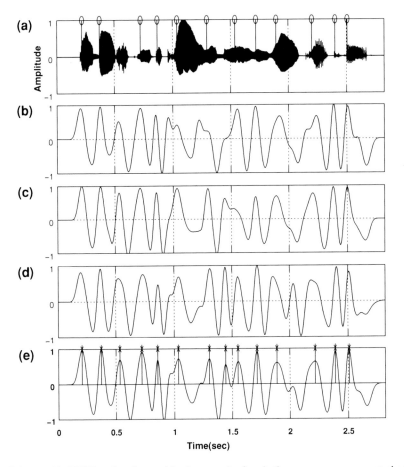

Fig. 2.4 Identified VOPs using the combined approach of excitation source energy, spectral peak energy, and modulation spectrum energy. **a** Speech signal with manually marked VOPs. **b** Evidence plot using excitation source energy. **c** Evidence plot using spectral peak energies. **d** Evidence plot using modulation spectrum energies. **e** Combined evidence plot

2.3.5 Pitch Synchronous Analysis

Analysis of the speech signal with respect to each pitch period is known as pitch synchronous analysis. In a normal block processing approach, physical decomposition of speech signal into the segments of length 20 ms is performed. In case of pitch synchronous analysis, logical decomposition of the speech signal, to cover one or multiple pitch cycles, is considered for feature extraction. This approach helps to derive logically related feature vectors. Variation in the speech signal between the consecutive pitch cycles is captured through pitch synchronous analysis.

Figure 2.7 shows the variation in spectral properties obtained from the consecutive pitch cycles of a voiced speech segment. From the gross observation, the spectra

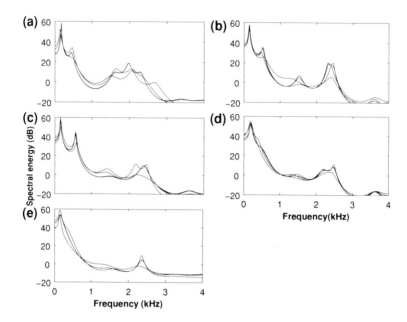

Fig. 2.5 Spectra of CV transition regions for 5 emotions. **a** Anger, **b** Fear, **c** Happiness, **d** Neutral and **e** Sadness

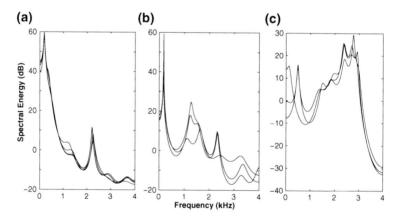

Fig. 2.6 Spectral properties of vowel, CV transition and consonant regions in the continuous speech. **a** Spectra for steady portion of the vowel, **b** Spectra for CV transition region and **c** Spectra for the consonant region

appear to be similar (close to each other), but at the finer level one can observe the variations in formant strengths and bandwidths in a successive pitch cycles. The sequence of finer variations may provide the desired emotion discrimination. In the literature 3–4 consecutive pitch cycles are considered for feature extraction [11, 12], but in this study, every pitch cycle of entire voiced region of the utterances is

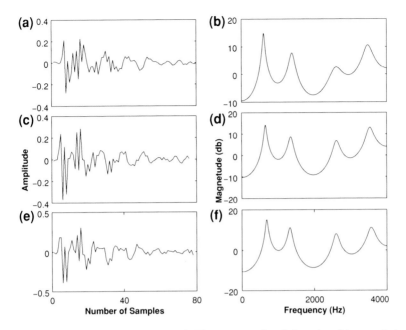

Fig. 2.7 Pitch synchronous analysis. **a, c** and **e** Three consecutive pitch cycles of the speech signal. **b, d** and **f** Corresponding spectra

independently processed for extracting the spectral features. Some of the important intuitions behind using pitch synchronous analysis of speech signals are as follows: The illogical approach associated with physical framing of the speech signal practiced in block processing, can be eliminated by carrying the analysis of speech signal within a pitch period. The assumption that a speech signal is stationary within the frame of 20 ms is not completely acceptable as both source and system are continuously varying with respect to time. In this work, pitch periods are marked using glottal closure instants (GCIs). A signal between two consecutive GCIs is treated as one pitch cycle. A zero frequency filter based method is used to determine the GCIs [13]. LPCCs, MFCCs and formant features are computed for each pitch cycle of a speech signal.

2.4 Classifiers

GMMs and AANNs are known to capture the general distribution of data points in the feature space and one can be used as an alternative to the other [14]. Two classifiers are used in this study, to mutually compare their emotion classification results.

2.4.1 Gaussian Mixture Models (GMM)

Gaussian Mixture Models (GMMs) are among the most statistically mature methods for clustering. A Gaussian mixture model is used as a classification tool in this task. They model the probability density function of observed variables using a multivariate Gaussian mixture density. Given a series of inputs, a GMM refines the weights of each distribution through the expectation-maximization algorithm. Mixture models are a type of density model, which comprise a number of component functions, usually Gaussian in nature. These component functions are combined to provide a multi-modal density. Mixture models are a semi-parametric alternative to non-parametric models and provide greater flexibility and precision in modeling the underlying statistics of sample data. They are able to smooth over gaps resulting from sparse sample data and provide tighter constraints in assigning object membership to cluster regions. Once a model is generated, conditional probabilities can be computed for test patterns (unknown data points). An expectation maximization (EM) algorithm is used for finding maximum likelihood estimates of parameters in probabilistic models, where the model depends on unobserved latent variables.

Expectation Maximization is an iterative method that alternates between performing an expectation (E) step, which computes an expectation of the log likelihood with respect to the current estimate of the distribution for the latent variables, and a maximization (M) step, which computes the parameters that maximize the expected log likelihood found on the E step. These parameters are then used to determine the distribution of the latent variables in the next E step.

The number of Gaussians in the mixture model is also known as the number of components. They indicate the number of clusters in which data points are to be distributed in order to cover local variations. In this work, one GMM is developed to capture the information about one emotion. Depending on the number of training data points, the number of components may be varied in each GMM. The presence of few components in a GMM, and training it with large number of data points may lead to more generalized clusters, failing to capture specific details related to each class. On the other hand over-fitting of the data points may happen, if too many components represent few data points. Obviously the complexity of the model increases, if they contain higher numbers of components. Therefore a tradeoff has to be reached between the complexity and the accuracy of the classification results required. In this work, GMMs are designed with 64 components and iterated 30 times to attain convergence. A diagonal covariance matrix is used to derive the model parameters.

2.4.2 Auto-Associative Neural Networks

AANN models are basically feed-forward neural network (FFNN) models, which try to map an input vector onto itself, and hence the name auto-association or identity

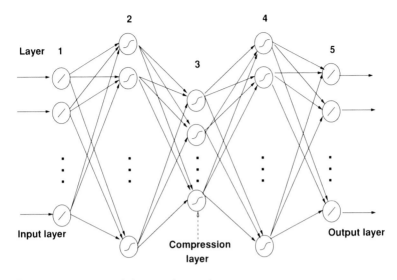

Fig. 2.8 Five layer auto-associative neural network

mapping [15, 16]. It consists of an input layer, an output layer and one or more hidden layers. The number of units in the input and output layers is equal to the dimension of the input feature vectors. The number of nodes in one of the hidden layers is less than the number of units in either the input or output layer. This hidden layer is also known as the dimension compression layer. The activation function of the units in the input and output layers is linear, whereas in case of hidden layers it is either linear or nonlinear. A five-layer AANN model with the structure shown in Fig. 2.8 is used in this study. The structure of network is generally represented by PL-QN-RN-QN-PL, where P, Q and R refers to integer values, L refers to linear units and N to nonlinear units. The integer value indicates the number of units present in that layer. The number of linear elements at the input layer indicates, the size of the feature vectors used for developing the models. The nonlinear units use $tanh(s)$ as the activation function, where s is the net input value of that unit. The structure of the network was determined empirically.

The performance of AANN models can be interpreted in different ways, depending on the application and the input data. If the data is a set of feature vectors in the feature space, then the performance of AANN models can be interpreted as linear or nonlinear principal component analysis (PCA) or capturing the distribution of input data [17–19]. On the other hand, if the AANN is presented directly with signal samples, such as LP residual signal, the network captures the implicit linear/nonlinear relations among the samples [20–22].

Determining the network structure is an optimization problem. At present there are no formal methods for determining the optimal structure of a neural network. The key factors that influence the neural network structure are learning ability of a network and capacity to generalize the acquired knowledge. From the available literature, it is observed that 5 layer symmetric neural networks, with three hidden layers have been used for different speech tasks. The first and the third hidden layers have more number of nodes than the input or output layer. The middle layer (also known as dimension compression layer) contains fewer units [23, 24]. In this type of network, generally the first and third hidden layers are expected to capture the local information among the feature vectors and the middle hidden layer is meant for capturing global information. Most of the existing studies [23–26] have used the 5 layer AANNs with the structure $N_1 L - N_2 N - N_3 N - N_2 N - N_1 L$, for their optimal performance. Here N_1, N_2, and N_3 indicate the number of units in the first, second and third layers respectively, of the symmetric 5-layer AANN. Usually N_2 and N_3 are derived experimentally, for achieving the best performance in the given task. From the existing studies, it is observed that N_2 is in the range of 1.3–2 times N_1 and N_3 is in the range of 0.2–0.6 times N_1. For designing the structure of the network, we have used the guidelines from the existing studies and experimented with few structures for finalizing the optimal structure. The performance of the network does not depend critically on the structure of the network [21, 27–29]. The number of units in the two hidden layers is guided by the heuristic arguments given above. All the input and output features are normalized to the range $[-1, +1]$ before presenting to the neural network. The back-propagation learning algorithm is used for adjusting the weights of the network to minimize the mean squared error [24].

2.5 Results and Discussion

In this work, 21 emotion recognition systems (ERS) are developed to study speech emotion recognition using different spectral features. In the beginning, emotion recognition systems are developed individually, using MFCCs, LPCCs, and formant features. Formant features alone have not given appreciably good emotion recognition performance, therefore, in the later stages, they are used in combination with the other features. Totally 5 sets of emotion recognition systems are developed as shown in Fig. 2.9. They are the ERSs developed using the spectral features derived from (a) the entire speech signal, (b) the vowel region, (c) the consonant region, (d) the CV transition region, and (e) pitch synchronous analysis. In each set, emotion recognition systems are developed using LPCCs, MFCCs, LPCCs+formant features, and MFCCs+formant features. In the following paragraphs, the emotion recognition performance of all individual emotion recognition systems, developed using Set3 of IITKGP-SESC, are discussed. Out of 10 speakers' speech data, the utterances of 8 speakers (4 male and 4 female) are used for training the ER models and the utterances of 2 (a male and a female) speakers are used for validating the trained models. Thirteen spectral features are extracted from a frame of 20 ms, with a shift of 5 ms. GMMs with 64 components are used to develop ERSs. The results of emotion recognition performance using session and text independent (Set1 and Set2) speech data are also given at the end of the chapter.

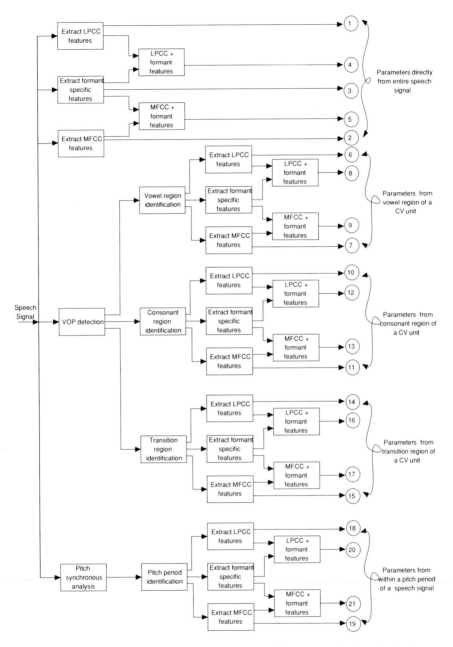

Fig. 2.9 Block diagram indicating various emotion recognition systems developed using the proposed spectral features, extracted from the entire speech utterance, sub-syllabic segments (vowel, consonant, and CV transition), and pitch synchronous analysis

Table 2.3 Emotion classification performance using the spectral features obtained from entire speech signal, adopting conventional block processing approach (ERSs 1–5)

Emotions	LPCCs (ERS1)	MFCCs (ERS2)	Formant features (ERS3)	LPCCs+ formant features (ERS4)	MFCCs+ formant features (ERS5)
	Emotion recognition performance in %				
Anger	53	57	33	60	63
Disgust	63	53	33	60	60
Fear	67	60	37	63	63
Happy	77	70	47	73	70
Neutral	80	77	66	83	77
Sadness	70	63	57	73	73
Sarcasm	73	70	64	77	60
Surprise	60	57	40	63	53
Average	68	63.38	47	69	68

ERS1 is developed using 13 LPCC features obtained from the entire speech signal using the normal block processing approach. ERS2 is developed using 13 MFCC features extracted frame wise from entire speech signal. ERS3 is developed using formant related features. These 13 formant related features (4 frequencies, 4 energy values, 4 bandwidth values and a slope), extracted per frame of 20 ms, are used to represent formant information. Their concatenation forms the 13 dimensional feature vector. The average emotion recognition performance using formant features is about 47 %. ERS4 and ERS5 are developed using the combination of 13 formant features along with 13 LPCCs and 13 MFCCs respectively. The dimension of the resulting feature vectors is 26. Table 2.3 shows the emotion recognition performance of ERS1–ERS5.

ERS6, ERS7, ERS8 and ERS9 are developed using the spectral features extracted from the vowel regions of the utterances [30]. LPCCs, MFCCs and formant features are extracted from 60 ms of speech signal, chosen from the steady portion of the vowel region of each syllable. Table 2.4 shows the performance of ERSs 6–9.

Similar to the emotion recognition systems developed for vowel regions (ERS6–ERS9), four systems are developed for consonant regions (ERS10–ERS13) and four systems are developed for CV transition regions (ERS14–ERS17) [30]. Tables 2.5 and 2.6 show the emotion recognition performance of ERSs developed using LPCC features, MFCC features, LPCCs+formant features and MFCCs+formant features, extracted from consonant and CV transition regions of the syllables, respectively.

Pitch synchronously extracted spectral features are used to capture the finer level spectral dynamics specific to the speech emotions. Therefore, spectral and formant features, extracted from each pitch period are used to develop the ERSs 18–21. Table 2.7 shows the emotion recognition performance using the spectral features extracted through pitch synchronous analysis.

Table 2.4 Emotion classification performance using the spectral features obtained from the vowel regions (ERSs 6–9)

Emotions	LPCCs (ERS6)	MFCCs (ERS7)	LPCCs+ formant features (ERS8)	MFCCs+ formant features (ERS9)
	Emotion recognition performance in %			
Anger	53	40	53	47
Disgust	50	43	53	40
Fear	50	50	57	43
Happy	47	53	50	60
Neutral	63	60	63	67
Sadness	60	57	67	50
Sarcastic	63	50	63	53
Surprise	53	47	57	47
Average	54.88	50	58	50.88

Table 2.5 Emotion classification performance using the spectral features obtained from the consonant regions (ERSs 10–13)

Emotions	LPCCs (ERS10)	MFCCs (ERS11)	LPCCs+ formant features (ERS12)	MFCCs+ formant features (ERS13)
	Emotion recognition performance in %			
Anger	37	33	40	37
Disgust	40	37	50	43
Fear	43	40	47	33
Happy	43	43	50	40
Neutral	47	50	57	53
Sadness	50	47	53	47
Sarcastic	50	53	53	47
Surprise	33	30	43	43
Average	42.88	41.63	49.13	42.88

The contribution of different features and different sub-syllabic regions toward specific emotions can be analyzed by observing individual emotion recognition performance of all ERSs. In general, combination of LPCCs and formants (using block processing and pitch synchronous analysis) are more discriminative with respect to all emotions. Specifically, the features from CV transition regions performed better in case of slow emotions like sadness and neutral. Happy is generally recognized well by most of the features. The results of the above studies show that emotion recognition performance using LPCCs is better than the results of MFCCs. LPCCs in the cases of the conventional block processing and pitch synchronous analysis

Table 2.6 Emotion classification performance using the spectral features obtained from the consonant-vowel transition regions (ERSs 14–17)

Emotions	LPCCs (ERS14)	MFCCs (ERS15)	LPCCs+ formant features (ERS16)	MFCCs+ formant features (ERS17)
	Emotion recognition performance in %			
Anger	47	57	53	57
Disgust	57	53	67	67
Fear	63	53	64	60
Happy	70	50	67	67
Neutral	77	67	77	80
Sadness	63	63	80	73
Sarcastic	67	67	73	70
Surprise	60	57	63	63
Average	63.13	58.38	68	67.13

Table 2.7 Emotion classification performance using pitch synchronously extracted spectral features (ERSs 18–21)

Emotions	LPCCs (ERS18)	MFCCs (ERS19)	LPCCs+ formant features (ERS20)	MFCCs+ formant features (ERS21)
	Emotion recognition performance in %			
Anger	57	50	60	60
Disgust	63	60	67	63
Fear	60	67	63	63
Happy	70	70	73	67
Neutral	80	77	77	77
Sadness	77	73	80	80
Sarcastic	73	63	70	77
Surprise	67	70	73	63
Average	68.38	66.25	70.38	68.75

have achieved highest emotion recognition of around 69 %. The reason for this may be that LPCCs mainly represent the speech production characteristics, by analyzing all frequency components in a uniform manner. The emotion specific information may be present across all the frequencies in a uniform way. The proposed formant features alone are not suitable to develop emotion recognition systems as their individual performance is poor. However, the combination of formant features with other spectral features has been proved to improve the recognition performance.

Emotion recognition performance using the spectral features from the consonant region is very poor. This may be due to poor representation of consonant regions by spectral features. The systems developed using only vowel regions have shown the

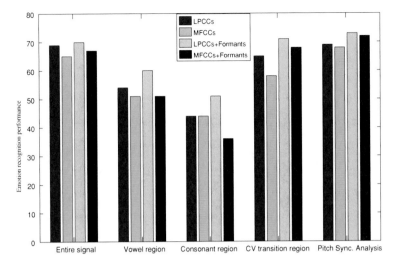

Fig. 2.10 Comparison of emotion recognition performance with respect to the entire speech signal, sub-syllabic speech segments, and pitch synchronous analysis, using proposed spectral features

average emotion recognition performance of around 58 %. The low performance of the spectral features in the vowel regions is due to the presence of redundant information. The special observation of the results presented in Tables 2.4, 2.5, 2.6 and 2.7 is that the CV transition regions contain more emotion specific information than vowel and consonant regions. The models developed using CV transition regions have achieved the performance as equal as that of a conventional block processing approach. It indicates that extracting the features from CV transition regions alone would be sufficient to recognize the underlying emotions. Processing only CV transition regions for emotion classification has two important advantages: (1) it is highly effective in the view of computational complexity and (2) achievement of almost similar performance due to selective processing of crucial information.

Compared to other spectral features, system features extracted from individual pitch cycles have performed well. The reason for this is that spectral characteristics are computed from each pitch cycle, with the intention of capturing finer spectral variations among the successive pitch cycles. It may be also noted that finer spectral variations are more emotion-specific than the frame-wise spectral information. The observations discussed above may be visualized from the bar graphs shown in Figs. 2.10 and 2.11. Figure 2.10 shows the emotion recognition performance of various spectral features extracted from the entire speech signal, sub-syllabic regions and pitch synchronous analysis. Figure 2.11 gives the emotion recognition performance by LPCC+formant features extracted from different speech regions and pitch synchronous analysis.

Different numbers of LPCCs/MFCCs are also explored for analyzing the emotion recognition performance. Table 2.8 indicates the average emotion recognition performance of 8 emotions using 8, 13 and 21 spectral features. It may be observed

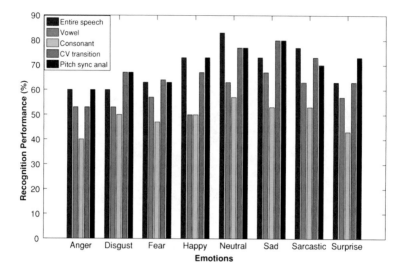

Fig. 2.11 Comparison of emotion recognition performance of the LPCCs+formant features, with respect to different emotions, obtained from entire speech signal, sub-syllabic regions, and pitch synchronous analysis

Table 2.8 Average emotion classification performance on IITKGP-SESC using different lengths of cepstral vectors

Methods & regions used for feature extraction	Features	No. of Ceps. Coeff.		
		8	13	21
		Recognition performance (in %)		
Block processing (Entire speech signal)	LPCCs	59	68	69
	MFCCs	55	63	65
	Formant features		47	
	LPCCs+Formants	58	69	70
	MFCCs+Formants	57	65	67
Vowel region	LPCCs	48	55	54
	MFCCs	42	50	51
	LPCCs+Formants	50	58	60
	MFCCs+Formants	43	51	51
Consonant region	LPCCs	35	43	44
	MFCCs	31	42	44
	LPCCs+Formants	38	49	51
	MFCCs+Formants	31	43	36

continued

Table 2.8 continued

Methods & regions used for feature extraction	Features	No. of Ceps. Coeff.		
		8	13	21
		Recognition performance (in %)		
CV-transition region	LPCCs	54	63	65
	MFCCs	50	58	58
	LPCCs+Formants	58	68	71
	MFCCs+Formants	55	67	68
Pitch synchronous analysis	LPCCs	56	68	69
	MFCCs	58	66	68
	LPCCs+Formants	61	70	73
	MFCCs+Formants	60	69	72

from Table 2.8 that most of the times, the systems developed using higher numbers of spectral features have performed slightly better than their counterparts developed using smaller numbers of spectral features. It may be due to the reason that higher order spectral features contain more specific information about paralinguistic aspects of the speech, such as speaker, rhythm, melody, timbre, emotion and so on [31].

Auto-associative neural networks are known for capturing the non-linear relations among the feature vectors [32]. AANNs are also capable of capturing the distribution properties as GMMs do [14]. So the best performance by GMMs, using 21 spectral features, is compared with the results of relevant AANNs. AANNs are used as the emotion classifiers to cross validate empirically the results obtained by GMMs. Table 2.9 shows the comparison of emotion recognition results of both GMMs and AANNs. AANN structures used for developing emotion models are also given in the last column of the table. From the results, it is observed that emotion recognition performance using GMMs is better than that of AANN models. This indicates that emotion specific information from the spectral features is better captured by GMMs than by the AANNs. The basic purpose of any emotion recognition system is to recognize the real life emotions with greater accuracy. Recognition of real emotions using the proposed sub-syllabic and pitch synchronous spectral features is discussed in Chap. 6. The proposed spectral features are also tested on internationally known Berlin emotion speech corpus (Emo-DB). The results obtained are on par with the results of our Indian database (IITKGP-SESC). Table 2.10 shows the comparison of emotion recognition results obtained using IITKGP-SESC and Emo-DB.

The results given in Table 2.10 are obtained using 21 spectral and 13 formant features. The emotion recognition systems are developed using GMMs. From the table, it is evident that the trends of emotion recognition performance using different spectral features on IITKGP-SESC and Emo-DB are almost the same.

So far we have carried out emotion recognition using the Set3 data set of IITKGP-SESC, which represents speaker and text independent emotion recognition. Along

Table 2.9 Average emotion classification performance of GMM and AANN models using spectral features on IITKGP-SESC

Features	Methods & regions for feature extr.	GMMs	AANNs	AANN structure
LPCCs	Block	69	63	21-45-10-45-21
MFCCs	processing	65	59	21-45-10-45-21
Formant features	(Entire	47	41	13-28-7-28-13
LPCCs+Formants	speech	70	61	34-60-15-60-34
MFCCs+Formants	signal)	67	60	34-60-15-60-34
LPCCs	Vowel	54	48	21-45-10-45-21
MFCCs	region	51	43	21-45-10-45-21
LPCCs+Formants		60	**57**	34-60-15-60-34
MFCCs+Formants		51	45	34-60-15-60-34
LPCCs	Consonant	44	39	21-45-10-45-21
MFCCs	region	44	37	21-45-10-45-21
LPCCs+Formants		51	46	34-60-15-60-34
MFCCs+Formants		36	33	34-60-15-60-34
LPCCs	CV-	65	57	21-45-10-45-21
MFCCs	transition	58	51	21-45-10-45-21
LPCCs+Formants	region	71	64	34-60-15-60-34
MFCCs+Formants		68	66	34-60-15-60-34
LPCCs	Pitch	69	64	21-45-10-45-21
MFCCs	synchronous	68	61	21-45-10-45-21
LPCCs+Formants	analysis	73	67	34-60-15-60-34
MFCCs+Formants		72	68	34-60-15-60-34

Table 2.10 Average emotion classification performance using the proposed spectral features on IITKGP-SESC and Emo-DB

Features	Methods & regions for feature extr.	IIT KGP-SESC	Emo-DB
LPCCs	Block	69	64
MFCCs	processing	65	63
Formant features	(Entire	47	41
LPCCs+Formants	speech	70	68
MFCCs+Formants	Signal)	67	63
LPCCs	Vowel	54	51
MFCCs	region	51	50
LPCCs+Formants		60	57
MFCCs+Formants		51	49

continued

Table 2.10 continued

Features	Methods & regions for feature extr.	IIT KGP-SESC	Emo-DB
LPCCs	Consonant	44	40
MFCCs	region	44	42
LPCCs+Formants		51	51
MFCCs+Formants		36	34
LPCCs	CV-	65	66
MFCCs	transition	58	57
LPCCs+Formants	region	71	69
MFCCs+Formants		68	64
LPCCs	Pitch	69	67
MFCCs	synchronous	68	65
LPCCs+Formants	analysis	73	72
MFCCs+Formants		72	70

Table 2.11 Average emotion classification performance using proposed spectral features on Set1, Set2, and Set3 datasets of IITKGP-SESC

Features	Methods & regions for feature extr.	IITKGP-SESC (rec.%)		
		Set1	Set2	Set3
LPCCs	Block	74	71	69
MFCCs	processing	67	65	65
Formant features	(Entire	53	49	47
LPCCs+Formants	speech	75	72	70
MFCCs+Formants	Signal)	71	68	67
LPCCs	Vowel	57	55	54
MFCCs	region	56	54	51
LPCCs+Formants		65	62	60
MFCCs+Formants		53	52	51
LPCCs	Consonant	47	45	44
MFCCs	region	47	44	44
LPCCs+Formants		55	53	51
MFCCs+Formants		40	38	36
LPCCs	CV-	68	66	65
MFCCs	transition	63	60	58
LPCCs+Formants	region	74	72	71
MFCCs+Formants		72	71	68
LPCCs	Pitch	75	73	69
MFCCs	synchronous	72	69	68
LPCCs+Formants	analysis	77	75	73
MFCCs+Formants		74	73	72

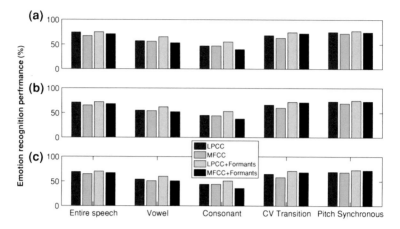

Fig. 2.12 Comparison of emotion recognition performance of different proposed spectral features with respect to the entire speech, sub-syllabic regions, and pitch synchronous analysis on Set1, Set2, and Set3 of IITKGP-SESC. **a** Emotion recognition performance using Set1, **b** Emotion recognition performance using Set2, and **c** Emotion recognition performance using Set3

with Set3, emotion recognition results of Set1 and Set2 are given in Table 2.11. Recognition performance of Set2 data set is 2 % higher than the results of Set3 data set. This is mainly due to the influence of speaker specific information during emotion classification. In the case of Set1, the recognition performance is about 4–5 % more than the results of Set3 and about 2–3 % more than the results of Set2. This improvement in emotion recognition is due to text and speaker specific information. The comparison of emotion recognition performance using proposed spectral features on Set1, Set2, and Set3 of IITKGP-SESC is given in the bar graph shown in Fig. 2.12.

2.6 Summary

In this chapter, spectral features derived from sub-syllabic regions and pitch synchronous analysis are proposed for recognizing the emotions from speech. IITKGP-SESC and Emo-DB are used to carry out the emotion classification using the proposed spectral features. LPCCs, MFCCs and formant features are used as features to represent vocal tract information. Spectral features derived from sub-syllabic regions are independently analyzed for classifying the emotions. It may be concluded from the results that the entire speech signal may not be necessary to recognize underlying emotions. Hence, the redundant information present in the steady region of the vowel may be exempted from feature extraction. The spectral features extracted from CV transition regions have achieved the emotion recognition performance, almost comparable with the performance obtained using the entire speech signal. Pitch synchronously extracted spectral features outperformed the other spectral features while recognizing the emotions. The combination of LPCCs and formant features also has demonstrated better emotion recognition performance. The studies conducted in this chapter indicate that spectral features contain more discriminating properties with respect to different emotions than excitation source features. In the literature, MFCCs are claimed to be robust features for majority of the speech tasks such as: speech recognition and synthesis. Surprisingly, in this study, LPCCs are found to be outperforming MFCCs while classifying the emotions. Formant features combined with basic spectral features have always improved the emotion recognition performance of the systems by a consistent margin of around 2–3 %. Two classification models, namely GMMs and AANNs, are used for developing emotion recognition systems. The majority of the results reported in this chapter are obtained using GMMs as they performed slightly better than AANNs.

References

1. D. Ververidis, C. Kotropoulos, Emotional speech recognition: Resources, features, and methods. SPC **48**, 1162–1181 (2006)
2. D. Neiberg, K. Elenius, K. Laskowski, Emotion recognition in spontaneous speech using GMMs, in *INTERSPEECH 2006—ICSLP*, (Pittsburgh, Pennsylvania), pp. 809–812, 17–19 Sept 2006
3. D. Bitouk, R. Verma, A. Nenkova, Class-level spectral features for emotion recognition, Speech Commun. (2010) (in Press)
4. S.G. Koolagudi, S. Maity, V.A. Kumar, S. Chakrabarti, K.S. Rao, IITKGP-SESC: Speech Database for Emotion Analysis. Communications in Computer and Information Science, JIIT University, Noida, India, Springer. ISSN: 1865–0929 ed., 17–19 Aug 2009
5. F. Burkhardt, A. Paeschke, M. Rolfes, W. Sendlmeier, B. Weiss, A database of german emotional speech, in *Interspeech*, Lissabon, 2005
6. L.R. Rabiner, B.H. Juang, *Fundamentals of Speech Recognition* (Prentice-Hall, Englewood Cliffs, New Jersy, 1993)
7. J. Chen, Y.A. Huang, Q. Li, K.K. Paliwal, Recognition of noisy speech using dynamic spectral subband centroids. IEEE Signal Process. Lett. **11**, 258–261 (2004)
8. S.V. Gangashetty, C.C. Sekhar, B. Yegnanarayana, Detection of vowel on set points in continuous speech using auto-associative neural network models, in *INTERSPEECH*, IEEE, 2004
9. K.S. Rao, B. Yegnanarayana, Duration modification using glottal closure instants and vowel onset points. Speech Commun. **51**, 1263–1269 (2009)
10. S.R.M. Prasanna, B.V.S. Reddy, P. Krishnamoorthy, Vowel onset point detection using source, spectral peaks, and modulation spectrum energies. IEEE Trans. Audio Speech Lang. Process. **17**, 556–565 (2009)
11. Y. Zeng, H. Wu, R. Gao, Pitch synchronous analysis method and fisher criterion based speaker identification, in *Third International Conference on Natural Computation*, vol. 2 (IEEE Computer Society, Washington DC, USA, 2007), pp. 691–695. ISBN: 0-7695-2875-9
12. H. Muta, T. Baer, K. Wagatsuma, T. Muraoka, H. Fukuda, Pitch synchronous analysis of hoarseness in running speech. J. Acoust. Soc. Am. **84**, 1292–1301 (1988)
13. K. Murty, B. Yegnanarayana, Epoch extraction from speech signals. IEEE Trans. Audio Speech Lang. Process. **16**, 1602–1613 (2008)
14. B. Yegnanarayana, S.P. Kishore, AANN an alternative to GMM for pattern recognition. Neural Networks **15**, 459–469 (2002)

15. B. Yegnanarayana, *Artificial Neural Networks* (Prentice-Hall, New Delhi, India, 1999)
16. S. Haykin, *Neural Networks: A Comprehensive Foundation* (Pearson Education Aisa, Inc., New Delhi, India, 1999)
17. K.I. Diamantaras, S.Y. Kung, *Principal Component Neural Networks: Theory and Applications* (Wiley, New York, 1996)
18. M.S. Ikbal, H. Misra, B. Yegnanarayana, Analysis of autoassociative mapping neural networks, in *Proceedings of the International Joint Conference on Neural Networks (IJCNN)*, (USA, 1999), pp. 854–858
19. S.P. Kishore, B. Yegnanarayana, Online text-independent speaker verification system using autoassociative neural network models, in *Proceedings of the International Joint Conference on Neural Networks (IJCNN)*, vol. 2 (Washington, DC, USA, 2001), pp. 1548–1553
20. A.V.N.S. Anjani, Autoassociate neural network models for processing degraded speech, Master's Thesis, MS Thesis, Department of Computer Science and Engineering, Indian Institute of Technology Madras, Chennai 600 036, India, 2000
21. K.S. Reddy, Source and system features for speaker recognition, Master's Thesis, MS Thesis, Department of Computer Science and Engineering, Indian Institute of Technology Madras, Chennai 600 036, India, 2004
22. C.S. Gupta, Significance of source features for speaker recognition, Master's Thesis, MS Thesis, Department of Computer Science and Engineering, Indian Institute of Technology Madras, Chennai 600 036, India, 2003
23. S. Desai, A. W. Black, B. Yegnanarayana, K. Prahallad, Spectral mapping using artificial neural networks for voice conversion. IEEE Trans. Audio Speech Lang. Process. **18**, 954–964 (2010)
24. K.S. Rao, B. Yegnanarayana, Intonation modeling for indian languages. Comput. Speech Lang. **23**, 240–256 (2009)
25. C.K. Mohan, B. Yegnanarayana, Classification of sport videos using edge-based features and autoassociative neural network models. Signal Image Video Process. **4**, 61–73 (2008). doi:10.1007/s11760-008-0097-9
26. L. Mary, B. Yegnanarayana, Autoassociative neural network models for language identification, in *International Conference on Intelligent Sensing and Information Processing*, IEEE, pp. 317–320, 24 Aug 2004. doi:10.1109/ICISIP.2004.1287674
27. B. Yegnanarayana, K.S. Reddy, S.P. Kishore, Source and system features for speaker recognition using aann models, in *IEEE International Conference on Acoustics, Speech, and Signal Processing*, (Salt Lake City, UT), May 2001
28. C.S. Gupta, S.R.M. Prasanna, B. Yegnanarayana, Autoassociative neural network models for online speaker verification using source features from vowels, in *International Joint Conference on Neural Networks*, (Honululu, Hawii, USA), May 2002
29. B. Yegnanarayana, K.S. Reddy, S.P. Kishore, Source and system features for speaker recognition using AANN models, in *Proceedings of the IEEE International Conference Acoustics, Speech, Signal Processing*, (Salt Lake City, Utah, USA), pp. 409–412, May 2001
30. S.G. Koolagudi, K.S. Rao, Emotion recognition from speech using sub-syllabic and pitch synchronous spectral features. Int. J. Speech Technol. **15**, 495–511 (2012). doi:10.1007/s10772-012-9150-8
31. O.M. Mubarak, E. Ambikairajah, J. Epps, Analysis of an mfcc-based audio indexing system for efficient coding of multimedia sources, in *The 8th International Symposium on Signal Processing and its Applications*, (Sydney, Australia), 28–31 Aug 2005
32. R.O. Duda, P.E. Hart, D.G. Stork, *Pattern Classification*, 2nd edn. (A Wiley-interscience Publications, Singapore, 2004)

Chapter 3
Robust Emotion Recognition using Sentence, Word and Syllable Level Prosodic Features

Abstract This chapter discuss about the use of prosodic information in discriminating the emotions. The motivation for exploring the prosodic features to recognize the emotions is illustrated using the gross statistics and time varying patterns of prosodic parameters. Prosodic correlates of speech such as energy, duration and pitch parameters are computed from the emotional utterances. Global prosodic features representing the gross statistics of prosody and local prosodic features representing the finer variations in prosody are introduced in this chapter for discriminating the emotions. These parameters are further extracted separately from different levels such as entire utterances, words and syllables. The analysis of contribution of emotional information by the initial, middle and final portions of sentences, words and syllables are studied. Use of support vector machines for classifying emotional utterances based on prosodic features has been demonstrated. Chapter ends with discussion on emotion recognition results and important conclusions.

3.1 Introduction

Chapter 2 has discussed vocal tract system features for recognizing emotions. From the results it is observed that vocal tract system features carry useful emotion specific information. Apart from vocal tract features, prosodic features also play a crucial role in recognizing the emotions from speech. In this chapter, extraction and use of various prosodic features are discussed for recognizing the emotions from speech utterances.

In the literature, prosodic features are treated as major correlates of vocal emotions. The effect of emotions on basic prosodic parameters such as pitch, energy, and duration is analyzed in several studies. Human beings also mostly use the prosodic cues for identifying the emotions present in day-to-day conversations. For instance pitch and energy values are high for active emotions like anger, whereas the same parameters are comparatively lower for the passive emotion like sadness. Duration used for expressing anger is shorter than the duration used for sadness. Existing

K. S. Rao and S. G. Koolagudi, *Robust Emotion Recognition using Spectral and Prosodic Features*, SpringerBriefs in Speech Technology,
DOI: 10.1007/978-1-4614-6360-3_3, © The Author(s) 2013

works on the use of prosody for emotion recognition have mainly focused on static prosodic values (mean, maximum, minimum and so on) for discriminating the emotions. However, the time varying dynamic nature of prosodic contours seems to be more emotion specific. According to the literature, there is little evidence on using the dynamic nature of prosodic contours for speech emotion recognition. Generally, emotion specific prosodic cues may not be present uniformly at all positions of the utterance. Some emotions like anger are dominantly perceivable from the initial portion of the utterances, whereas surprise is dominantly expressed at the final part of the utterance. In this Chapter, prosodic features at utterance, word, and syllable levels are analyzed, to study the contribution of these parts toward speech emotion recognition. We also plan to study the emotion specific information present at the different portions (initial, middle and final) of the utterances. The study on these issues are carried out by performing the following tasks: (1) Analysis of speech emotions using static prosodic features, (2) Investigating dynamic prosodic features at the utterance, word, and syllable levels, for discriminating the emotions, and (3) Combination of measures due to static (global) and dynamic (local) prosodic features at different levels to recognize the emotions. Support vector machine (SVM) models are used for developing the emotion recognition models for discriminating the emotions. Statistical parameters like mean, minimum, maximum, standard deviation, derived from the sequence of prosodic values, are known as global or static features. The parameters representing temporal dynamics of the prosodic contours are known as local or dynamic features. In this work, the terms, static and global, dynamic and local are used as synonyms.

The rest of the Chapter is organized as follows: Sect. 3.2 discusses the importance of prosodic features in classifying the speech emotions. Section 3.3 briefly mentions the motivation behind this study. Section 3.4 explains the details of extraction of static and dynamic prosodic features from various segments of speech utterances. Evaluation of the developed emotion recognition models and their performance are discussed in Sect. 3.5. A brief summary of the present work is given in Sect. 3.6.

3.2 Prosodic Features: Importance in Emotion Recognition

Normally human beings use dynamics of long term speech features like energy profile, intonation pattern, duration variations and formant tracks, to process and perceive the emotional content from the speech utterances. This might be the main reason for the extensive use of prosodic features by most of the research community. However, many times humans tend to get confused while distinguishing the emotions that share similar acoustical and prosodic properties. In real situations, humans are helped by linguistic, contextual and other modalities like facial expressions, while interpreting the emotions from the speech.

For analyzing emotion specific prosodic characteristics, single male and female emotional utterances of IITKGP-SESC are considered. Means of the distribution of the prosodic features are computed and used for classifying eight emotions. Table 3.1

Table 3.1 Mean values of the prosodic parameters for each of the emotions of IITKGP-SESC

Emotion	Male artist			Female artist		
	Duration (Sec)	Pitch (Hz)	Energy	Duration (Sec)	Pitch (Hz)	Energy
Anger	1.76	195.60	115.60	1.80	301.67	57.42
Disgust	1.62	188.05	73.42	1.67	308.62	54.01
Fear	1.79	210.70	147.31	1.89	312.07	76.65
Happiness	2.03	198.30	81.12	2.09	287.78	40.02
Neutral	1.93	184.37	83.13	2.04	267.13	40.89
Sadness	2.09	204.00	108.12	2.13	294.33	40.36
Sarcasm	2.16	188.44	98.57	2.20	301.11	34.20
Surprise	2.05	215.75	202.06	2.09	300.10	41.49

Table 3.2 Emotion classification performance using prosodic features

	Anger	Disgust	Fear	Happiness	Neutral	Sadness	Sarcasm	Surprise
Male, average: 45.38								
Anger	37	0	10	23	13	0	0	17
Disgust	0	40	0	3	20	27	7	3
Fear	17	0	63	0	13	0	0	7
Happiness	33	0	10	37	10	0	0	10
Neutral	0	10	0	0	53	27	10	0
sadness	0	0	0	0	23	60	17	0
Sarcasm	0	27	0	0	10	20	43	0
Surprise	23	27	0	0	20	0	0	30
Female, average: 50.88								
Anger	43	0	10	17	13	0	0	17
Disgust	0	47	0	0	3	0	27	23
Fear	13	0	67	10	3	0	0	7
Happiness	10	0	13	57	10	0	0	10
Neutral	0	17	0	0	50	23	0	10
sadness	0	7	0	0	33	60	0	0
Sarcasm	0	33	0	0	17	0	43	7
Surprise	3	37	10	0	0	10	0	40

Ang. Anger, *Dis.* Disgust, *H* Happiness, *Neu.* Neutral, *Sar.* Sarcasm, *Sur.* Surprise

shows the mean values of the basic prosodic parameters of different emotions for both male and female speakers.

The mean duration is calculated by averaging the durations of all utterances. Mean pitch is computed by averaging the frame level pitch values for all utterances. Energy is an average of the frame level energies calculated for each utterance. Frames of size 20 ms and a shift of 10 ms are used for the above calculations. Though this statistical analysis of prosody toward emotion is very simple, it gives a clear insight of emotion specific knowledge present in the prosodic features. Table 3.2 shows the emotion recognition results based on the above prosodic parameters. Here a

Table 3.3 Generic trends of prosodic behavior of male and female utterances for 8 different emotions

Emotion	Male artist						Female artist					
	Dur.	Ran.of dur.	Pitch	Ran.of pit.	Eng.	Ran.of eng.	Dur.	Ran.of dur.	Pitch	Ran.of pit.	Eng.	Ran.of eng.
Anger	<	>	=	<	=	=	<	>	>	=	=	=
Disgust	≪	<	<	<	≪	≪	≪	<	≫	<	<	<
Fear	=	<	>	=	≫	=	=	<	≫	≫	≫	≫
Happiness	>	<	=	=	<	=	>	=	=	=	<	=
Neutral	=	=	≪	=	<	=	=	≫	≪	<	<	≪
Sadness	>	=	=	=	>	≫	>	=	=	<	<	=
Sarcasm	≫	≫	<	≪	≪	<	≫	≪	>	>	≪	<
Surprise	>	≪	≫	≫	=	<	>	<	>	≫	<	<

≪—Very low, <—Low, =—Medium, >—High, ≫—Very high
Dur. Duration, *Ran.Dur.* Range of duration, *Ran.Pit.* Range of pitch, *Eng.* Energy, *Ran.Eng.* Range of energy

simple Euclidean distance measure is used to classify eight emotions. An average emotion recognition performance of around 45 and 51 % is observed for male and female speech respectively. From the results, it may be observed that there are mis-classifications among high arousal emotions like anger, happiness, and fear. Similar observations with respect to slow emotions such as disgust, sadness, and neutral are also seen. Most of the mis-classifications are biased toward neutral. Emotions expressed by female speakers are recognized fairly better than the emotions of male speakers.

From the first and second order statistics (mean and variance) of the prosodic parameters of one male and one female speaker, a qualitative analysis is presented in Table 3.3. From the symbols given in Table 3.3, it may be noted that for female speech, out of eight emotions, happiness, sadness, and sarcasm are with larger utterance level duration, whereas the duration for anger and disgust is less. Pitch values are very high for fear and disgust. Energy is observed to be very high for fear. It is interesting to note that these trends are not common between the genders. It indicates that the emotion expression cues, in the case of males and females, are slightly different. Table 3.3 represents the overall trend of the prosodic features related to the emotions expressed in the database IITKGP-SESC. The prosodic trends given in Table 3.3 are obtained based on the mean values of individual prosodic parameters computed over individual (local mean) and all emotions (global mean). A range with respect to overall mean (global mean) is fixed to qualitatively decide the trend of prosodic parameters for given emotions. Table 3.4 gives the details of the ranges of different prosodic parameters for deriving the trends. For example, to determine the trend of pitch with respect to anger, initially, the average pitch value of all the sentences of all the emotions (global mean) present in the database is computed. Later the average pitch of the anger sentences (local mean) is computed. The deviation of the pitch of anger sentences (local mean) from overall average pitch (global mean) helps to decide

Table 3.4 Static ranges for qualitative description of prosodic parameters for different emotions, used in Table 3.3.

Prosodic parameters	Very less	Less	Medium	High	Very high
Duration	$LM < 0.5GM$	$0.5GM \leq$ $LM <$ $0.9GM$	$0.9GM \leq$ $LM <$ $1.1GM$	$1.1GM \leq$ $LM <$ $1.5GM$	$1.5GM < LM$
Pitch and energy	$LM < 0.9GM$	$0.9GM \leq$ $LM <$ $0.95GM$	$0.95GM \leq$ $LM <$ $1.05GM$	$1.05GM \leq$ $LM <$ $1.1GM$	$1.1GM < LM$

the trend of pitch for anger. Similarly, trends for duration and energy are determined by computing their global and local mean values. In Table 3.4, LM represents local mean (Mean of particular prosodic parameter for the specific emotion) and GM represents global mean (Mean of particular prosodic parameter for all the emotion).

3.3 Motivation

From Table 3.1, it may be observed that the average static prosodic values such as energy, pitch, and duration are distinguishable for different emotions. Similarly, the time varying dynamics in the prosodic contours also represent emotion specific information. Figure 3.1 shows the dynamics in prosodic contours for different emotions. Obviously, there are inherent overlaps among these static and dynamic prosodic values with respect to the emotions. In the literature, several references are observed about using static prosodic features for speech emotion recognition. However, time dependent prosody variations may be used as the discrimination strategy, where static prosodic properties of different emotions show high overlap. Figure 3.1 shows three subplots indicating the (a) duration patterns of the sequence of syllables, (b) energy contours and (c) pitch contours of an utterance "*mAtA aur pitA kA Adar karnA chAhie*" in five different emotions. From the subplot indicating the duration patterns, one can observe the common trend of durations for all emotions. However, the trends also indicate that, for some emotions such as fear and happiness, the durations of the initial syllables of the utterance are longer, for happiness and neutral emotions middle syllables of the utterance seem to be longer, and the final syllables of the utterance seem to be longer for fear and anger (see Fig. 3.1a). From the energy plots, it is observed that the utterance with anger emotion has highest energy for the entire duration. Next to the anger emotion, fear and happiness show somewhat more energy than the other two emotions. The dynamics of energy contours can be used to discriminate fear and happiness (see Fig. 3.1b). It is observed from Fig. 3.1c that anger, happiness and neutral have somewhat higher pitch values, than the other two emotions. Using the dynamics (changes of prosodic values with respect to time) of pitch contours, easy discrimination is possible between anger, happiness and neutral

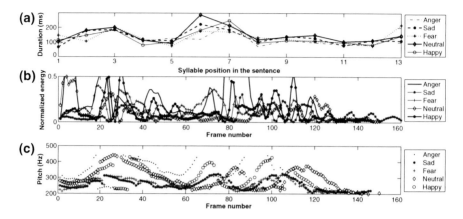

Fig. 3.1 **a** Duration patterns for the sequence of syllables, **b** energy contours, and **c** pitch contours in different emotions for the utterance *"mAtA aur pitA kA Adar karnA chAhie"*

emotions, even though they have similar average values. Thus, Fig. 3.1 provides the basic motivation to explore the dynamics of prosodic features for discriminating the emotions.

By observing pitch contours from Fig. 3.1c, it may be noted that for a given data, initial portions of the plots (the sequence of 20 pitch values) do not carry uniform discriminative information. Static features are almost the same for happiness and neutral. However, static features may be used to distinguish anger, sadness, fear emotions as their static pitch values are spread widely between 250 and 300 Hz. Similarly dynamic features are almost the same for all emotions except for fear. One may observe the initial decreasing and gradual rising trends of pitch contours for anger, happiness, neutral, and sadness emotions, whereas for the fear pitch contour starts with a rising trend. Similar local discriminative properties may also be observed in the case of energy and duration profiles from the initial, middle and final parts of the utterances. This phenomenon indicates that it may be sometimes difficult to classify the emotions based on either global or local prosodic trends derived from the entire utterance. Therefore, in this work, we intend to explore the static (global) and dynamic (local) prosodic features, along with their combination for speech emotion recognition at different levels (utterance, words, and syllables) and positions (initial, middle, and final).

3.4 Extraction of Global and Local Prosodic Features

In this Chapter, emotion recognition (ER) systems are developed using local and global prosodic features, extracted from sentence, word and syllable levels. Word and syllable boundaries are identified using vowel onset points (VOPs) as the anchor points. Detailed explanation of VOPs and their detection from speech signal are discussed in Sect. 2.3 of Chap. 2.

3.4.1 Sentence Level Features

Sentence level static and dynamic prosodic features are derived by considering the entire sentence as a unit for feature extraction. Pitch contours are extracted using the zero frequency filter based method [1]. Zero frequency filter method determines the instants of significant excitation (epochs) present in the voiced regions of speech signal. Voiced regions are determined using frame level energy and periodicity. In the unvoiced region, the concept of pitch is not valid, hence pitch values are considered as zero for each interval of 10 ms. In the voiced region pitch is determined using epoch intervals. The time interval between successive epochs is known as the epoch interval. Reciprocal of the epoch interval is considered as the pitch at that instant of time. The energy contour of a speech signal is derived from the sequence of frame energies. Frame energies are computed by summing the squared sample amplitudes within a frame. Fourteen (2-duration, 6-pitch, 6-energy) prosodic parameters are identified to represent duration, pitch and energy components of global prosodic features. Normalized syllable and pause durations are considered as two duration parameters. Normalized syllable duration is computed as

$$ND_{syl} = \frac{D_s - D_p}{N_{syl}}$$

and normalized pause duration is computed as

$$ND_p = \frac{D_p}{D_s}$$

where ND_{syl} is normalized syllable duration, D_s is Sentence duration, D_p is pause duration, N_{syl} is the number of syllables, and ND_p is normalized pause duration

Six pitch values are derived from the sentence level pitch contour. Those values are minimum, maximum, mean, standard deviation, median, and contour slope. The slopes of pitch and energy contours are determined by using the middle pitch and energy values of the first and the last words. These fourteen values are concatenated in the order duration, pitch, and energy to form a feature vector that represents global prosody.

Local prosodic features are intended to capture the variations in the prosodic contours with respect to time. Therefore, the feature vector is expected to retain the natural dynamics of the prosody. In this regard, resampled energy and pitch contours are used to represent the feature vectors for local prosody. The dimension of pitch and energy contours is chosen to be 25, after evaluating the emotion recognition performance with 100, 50, 25, and 10 values. The recognition performance with

25 dimensional feature vectors is slightly better than the feature vectors with other dimensions. Here, the dimension 25 for pitch and energy contours is not crucial. The reduced size of the pitch and energy contours has to be chosen so that the dynamics of the original contours are retained in their resampled versions. The basic reasons for reducing the dimensionality of the original pitch and energy contours are (1) the need for the fixed dimensional input feature vectors for developing the SVM models and (2) the number of feature vectors required for training the classifier has to be proportional to the size of the feature vector to avoid the curse of dimensionality (The need of number of feature vectors grows exponentially as the dimensionality of feature vector increases. Therefore always there should be a proportion between the number of available feature vectors and their dimensionality). The local duration pattern is represented by the sequence of normalized syllable durations. Here the syllable durations are determined using the time interval between successive VOPs [2]. The length of duration contour is proportional to the number of syllables present in the sentence, which leads to feature vectors of unequal lengths. To obtain the feature vectors of equal length, the length of duration vector is fixed to be 18 (the maximum number of syllables present in the longest utterance of IITKGP-SESC). The length for shorter utterances is compensated by zero padding.

3.4.2 Word and Syllable Level Features

The global and local prosodic features extracted from words and syllables help to analyze the contribution of different segments (sentences, words, and syllables) and their positions (initial, middle, and final), in the utterance toward emotion recognition. Word and syllable boundaries are determined automatically, using vowel onset points [3, 4]. Before extracting the features, the words in all the utterances of the database are divided into three groups namely initial, middle, and final words. Similarly, the syllables within each group of words are also classified as initial, middle, and final syllables. While categorizing the words, the length of the words and number of words in an utterance are taken into consideration. Length of words is measured in terms of number of syllables. If there are more than 3 words in the utterance and the first word is monosyllabic, then the first 2 words are grouped as initial words. This is because monosyllabic words may not be sufficient to capture emotion specific information. Many times monosyllabic words are not sufficient for the speaker to clearly express specific emotion. The scheme of grouping of words and syllables into the above mentioned three groups is given in Table 3.5. This table contains word and syllable grouping details for the 15 sentences of IITKGP-SESC. For instance, grouping of words in the case of (S1, S8, S9) and (S5, S11) is straight forward as there are either 3 or 6 words in the sentences. In the case of S2, out of 5, two words are grouped as the initial words, as the first word of the sentence is monosyllabic in nature. The last word, which contains 4 syllables, is treated as the final word, and the remaining two words are considered as the middle words. Similarly in the case of S3, the first word is considered as the initial word, as it contains 4 syllables. On the basis of production

Table 3.5 Linguistic details of text prompts of IITKGP-SESC: Scheme for groping of words and syllables while extracting prosodic parameters

Sen. ♯	♯ words	♯ Syllables in the word sequence	♯ syl.	♯ ini. wds.	♯ mid. wds.	♯ fin. wds.	♯ syl. in initial words	♯ syl. in middle words	♯ syl. in final words
S1	3	6 + 4 + 3	13	1	1	1	2 + 2 + 2	1 + 2 + 1	1 + 1 + 1
S2	5	1 + 2 + 2 + 4 + 4	13	2	2	1	2 + 0 + 1	2 + 2 + 2	1 + 2 + 1
S3	5	4 + 2 + 3 + 4 + 3	16	1	2	2	1 + 2 + 1	2 + 1 + 2	2 + 3 + 2
S4	4	4 + 4 + 3 + 3	14	1	1	2	1 + 2 + 1	1 + 2 + 1	2 + 2 + 2
S5	6	1 + 2 + 2 + 3 + 2 + 3	13	2	2	2	2 + 0 + 1	2 + 1 + 2	2 + 1 + 2
S6	5	4 + 2 + 5 + 3 + 3	17	2	1	2	2 + 2 + 2	1 + 3 + 1	2 + 2 + 2
S7	5	2 + 5 + 2 + 3 + 2	14	2	2	1	2 + 3 + 2	2 + 1 + 2	1 + 0 + 1
S8	3	3 + 4 + 4	11	1	1	1	1 + 1 + 1	1 + 2 + 1	1 + 2 + 1
S9	3	5 + 3 + 3	11	1	1	1	1 + 3 + 1	1 + 1 + 1	1 + 1 + 1
S10	5	1 + 2 + 6 + 3 + 2	14	2	1	2	2 + 0 + 1	1 + 4 + 1	2 + 1 + 2
S11	6	2 + 5 + 4 + 1 + 3 + 3	18	2	2	2	2 + 3 + 2	2 + 2 + 1	2 + 2 + 2
S12	4	2 + 2 + 4 + 4	12	2	1	1	2 + 0 + 2	1 + 2 + 1	1 + 2 + 1
S13	5	2 + 3 + 4 + 3 + 5	17	2	2	1	2 + 1 + 2	2 + 3 + 2	1 + 3 + 1
S14	4	3 + 2 + 3 + 3	11	1	2	1	1 + 1 + 1	2 + 1 + 2	1 + 1 + 1
S15	4	2 + 3 + 3 + 3	11	1	2	1	1 + 0 + 1	2 + 2 + 2	1 + 1 + 1

Fin. Final, *Ini.* Initial, *Mid.* Middle, *No.* Number, *Sen.* Sentences, *Syl.* Syllables, *Wds.* Words

and co-articulation constraints, words in each group are divided into initial, middle, and final syllables. The last 3 columns of Table 3.5 indicate the number of initial, middle, and final syllables present in initial, middle, and final words. Here the syllable division is carried out using the following principle. (a) If the word contains more than 2 syllables, then the first syllable of the word is considered as the initial syllable, the last syllable of the word is considered as the final syllable, and the remaining syllables are treated as the middle syllables. (b) If the word contains 2 syllables, then they are treated as the initial and final syllables. (c) If the word consists of a single syllable, then that syllable is treated as the initial syllable. The English transcriptions of the text prompts of the Telugu database (IITKGP-SESC) are given in Table 3.6. The unicode set for Telugu alphabet is available at [5].

The process of extracting word level global and local prosodic features is similar to the method of extracting utterance level global and local prosodic features. The length of the feature vectors for word level global prosodic features is kept as 13 (1-duration, 6-pitch, and 6-energy). Here, the parameter *normalized pause duration* is not included as the feature, since only one or two words are used for feature extraction. Slopes of the pitch and energy contours are computed by considering the first and last syllables of the specific words. The length of the feature vectors for word level local prosodic features is fixed to be 15 for pitch and energy. This is derived by re-sampling the original prosody contours obtained over the words. The length of local duration vector is fixed at 6, which is equal to the maximum number of syllables in a word of IITKGP-SESC. The length of the local duration vector, at

Table 3.6 English transcriptions of the Telugu text prompts of IITKGP-SESC

Sentence identity	Text prompts
S1	thallidhandrulanu gauravincha valenu
S2	mI kOsam chAlA sEpatnimchi chUsthunnAmu
S3	samAjamlo prathi okkaru chadhuvuko valenu
S4	ellappudu sathyamune paluka valenu
S5	I rOju nEnu tenali vellu chunnAnu
S6	kOpamunu vIdi sahanamunu pAtincha valenu
S7	anni dAnamulalo vidyA dAnamu minnA
S8	uchitha salahAlu ivvarAdhu
S9	dongathanamu cheyutA nEramu
S10	I rOju vAthAvaranamu podigA undhi
S11	dEsa vAsulandharu samaikhyAthA tho melaga valenu
S12	mana rAshtra rAjadhAni hyderAbAd
S13	sangha vidhrOha sekthulaku Ashrayam kalpincharAdhu
S14	thelupu rangu shAnthiki chihnamu
S15	gangA jalamu pavithra mainadhi

the syllable level, is fixed at 4, which is equal to the maximum number of syllables in any group, as shown in Table 3.5.

Out of 10 speakers of IITKGP-SESC, the speech utterances of eight speakers (4 male and 4 female) are used to train the emotion recognition models. Validation of the trained models is done using remaining 2 speakers' (1 male and 1 female) speech data. The details of the speech corpus, IITKGP-SESC, are given in Sect. 2.2 of Chap. 2. The description of development of emotion recognition models and their verification is discussed in the next Section.

3.5 Results and Discussion

Emotion recognition systems are separately developed for sentence, word, and syllable level global and local level prosodic features. The combination of global and local level features is also explored to study emotion recognition (ER).

3.5.1 Emotion Recognition Systems using Sentence Level Prosodic Features

In this work, we have considered 8 emotions of IITKGP-SESC, for studying the role of global and local prosodic features in recognizing speech emotions. SVMs are used to develop emotion recognition models. Each SVM is trained with positive and

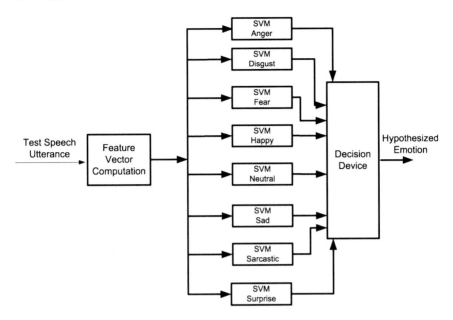

Fig. 3.2 General block architecture of an emotion recognition system using SVMs

negative examples. Positive feature vectors are derived from the utterances of the intended emotion, and negative feature vectors are derived from the utterances of all other emotions. Therefore, 8 SVMs are developed to represent 8 emotions. The basic block diagram of the ER system developed using SVMs is shown in Fig. 3.2. For evaluating the performance of the ER systems, the feature vectors are derived from the test utterances and are given as inputs to all 8 trained emotion models. The output of each model is given to the decision module, where the category of the emotion is hypothesized based on the highest evidence among the 8 emotion models.

For analyzing the effect of global and local prosodic features on emotion recognition performance, separate models are developed using global and local prosodic features [6]. The overall emotion recognition system consisting of combination of measures from the global and local prosodic features is shown in Fig. 3.3.

The emotion recognition system based on global prosodic features consists of 8 emotion models, developed by using 14-dimensional feature vectors (duration parameters-2, pitch parameters-6, energy parameters-6). Emotion recognition performance of the models using global prosodic features is given in Table 3.7. Fear and neutral are recognized with the highest rate of 67 %, whereas happiness utterances are identified with only 14 % of accuracy. It is difficult to attain high performance, while classifying the underlying speech emotions using only static prosodic features. This is mainly due to the overlap of static prosodic features of different emotions. For instance, it is difficult to discriminate pairs like fear and anger, sarcasm and disgust using global prosodic features. Utterances of all 8 emotions are mis-classified

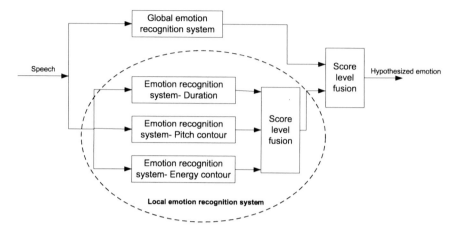

Fig. 3.3 Emotion recognition system using sentence level global and local prosodic features

Table 3.7 Emotion classification performance using global prosodic features computed over entire utterances

Emotions	Emotion recognition performance in %							
	Ang.	Dis.	Fear	Hap.	Neu.	Sad	Sar.	Sur.
Anger	28	17	23	3	13	13	3	0
Disgust	7	47	0	0	3	10	33	0
Fear	7	0	67	7	0	10	0	9
Happiness	3	0	7	14	43	3	10	20
Neutral	0	0	7	17	67	0	3	6
Sadness	7	3	17	17	0	40	13	3
Sarcasm	0	10	0	13	20	3	44	10
Surprise	7	0	17	13	3	3	13	44

Average recognition performance: 43.75

Ang. Anger, *Dis.* Disgust, *Hap.* Happiness, *Neu.* Neutral, *Sar.* Sarcasm, *Sur.* Surprise

as either neutral, fear, or happiness. The mis-classification due to static prosodic features may be reduced by employing dynamic prosodic features for classification. Therefore, use of dynamic nature of prosody contours, captured through local prosodic features, is explored in this work for speech emotion recognition.

To study the relevance of individual local prosodic features in emotion recognition, three separate ER systems corresponding to sentence level duration, intonation and energy patterns are developed to capture local emotion specific information. Score level combination of these individual local prosodic systems is performed to obtain overall emotion recognition performance due to all local utterance level features. Emotion recognition performance using individual local prosodic features and their score level combination is given in Table 3.8. The average emotion recognition performance due to individual local prosodic features is well above the performance of global prosodic features. The information of pitch dynamics has the highest dis-

Table 3.8 Emotion classification performance using local prosodic features computed over entire sentences

Emotions	Ang.	Dis.	Fear	Hap.	Neu.	Sad	Sar.	Sur.
Duration, average emotion recognition: 48.75								
Anger	30	20	7	3	23	3	7	7
Disgust	7	67	3	0	10	3	10	0
Fear	7	7	53	0	10	10	7	6
Happiness	17	7	10	30	3	13	7	13
Neutral	7	3	3	3	57	21	3	3
Sadness	3	7	23	7	20	30	10	0
Sarcasm	0	13	4	10	0	0	73	0
Surprise	7	3	14	10	3	10	3	50
Pitch, average emotion recognition: 53.75								
Anger	27	43	7	0	3	3	17	0
Disgust	10	60	10	0	0	3	7	0
Fear	3	13	43	7	0	10	7	17
Happiness	4	7	13	40	3	13	13	7
Neutral	3	7	0	3	80	7	0	0
Sadness	3	10	7	7	10	57	6	0
Sarcasm	0	0	7	3	0	10	63	17
Surprise	0	0	7	10	0	3	20	60
Energy, average emotion recognition: 48								
Anger	43	37	7	0	7	0	6	0
Disgust	27	37	0	0	13	0	20	3
Fear	0	0	57	7	10	13	0	13
Happiness	7	7	10	43	17	7	7	2
Neutral	20	3	3	10	47	10	3	4
Sadness	0	3	13	17	17	40	10	0
Sarcasm	0	10	3	0	0	0	80	7
Surprise	0	0	37	13	3	0	10	37
Duration + Pitch + Energy, Average: 64.38								
Anger	40	4	23	27	3	0	3	0
Disgust	13	73	0	0	4	0	10	0
Fear	3	0	63	10	0	7	0	17
Happiness	7	0	10	57	3	13	3	7
Neutral	0	7	7	3	73	7	0	3
Sadness	0	7	13	7	0	63	10	0
Sarcasm	0	10	0	0	0	7	83	0
Surprise	3	0	17	10	0	0	7	63

Ang. Anger, *Dis.* Disgust, *Hap.* Happiness, *Neu.* Neutral, *Sar.* Sarcasm, and *Sur.* Surprise

crimination of about 54 %. Energy and duration dynamic features have also achieved a recognition performance around 48 %. From the results, it is observed that local prosodic features play a major role in discriminating the emotions. Score level combination of energy, pitch and duration features further improved the emotion recognition performance up to around 65 %. Measures of emotion recognition models developed using global and local prosodic features are combined for improving the

Table 3.9 Emotion classification performance using the combination of local and global prosodic features computed from entire utterances

Emotions	Emotion recognition performance in %							
	Anger	Disgust	Fear	Happiness	Neutral	Sadness	Sarcasm	Surprise
Anger	47	40	3	0	3	0	7	0
Disgust	10	63	0	0	7	10	10	0
Fear	7	0	60	3	0	13	0	17
Happiness	10	0	7	53	20	0	3	7
Neutral	0	0	3	7	74	13	3	0
Sadness	0	0	17	3	0	77	3	0
Sarcasm	0	10	0	0	3	0	84	3
Surprise	0	0	10	17	0	0	6	67

Average recognition performance: 65.63 %

Ang. Anger, *Dis.* Disgust, *Hap.* Happiness, *Neu.* Neutral, *Sar.* Sarcasm, *Sur.* Surprise

performance further. Table 3.9 shows the recognition performance of the emotion recognition system developed by combining the measures from global and local prosodic features.

The average emotion recognition performance after combining the global and local prosodic features is observed to be about 65.63 %. There is no considerable improvement in the emotion recognition rate, by combining the measures from global and local prosodic features. This indicates that the emotion discriminative properties of global prosodic feature are not complementary to those of local features. Therefore, local prosodic features alone would be sufficient to perform speech emotion recognition. The comparison of recognition performance in case of each emotion, with respect to the global, local and their combination of features is shown in Fig. 3.4. It may be observed from the figure that anger, neutral, sadness, and surprise have achieved better discrimination using the combination of global and local prosodic features. Local prosodic features play an important role in the discrimination of disgust, happiness, and sarcasm. Fear is recognized well by using global prosodic features.

3.5.2 Emotion Recognition Systems using Word Level Prosodic Features

In general, while expressing emotions, different emotions appear to be effective at different parts of the utterances. For example, anger and happiness show their characteristics mainly at the beginning of the utterance. Effective expression of fear and disgust may be observed in the final part of the utterance. Based on this intuitive hypothesis, initial, middle, and final portions of the utterances are analyzed separately for capturing the emotion specific information. To analyze characteristics of emotions at different parts of the utterance, each utterance of IITKGP-SESC, is divided into

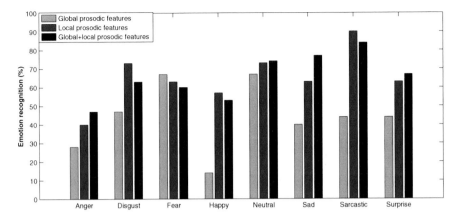

Fig. 3.4 Comparison of emotion recognition performance using utterance level global, local, and global+local prosodic features

3 parts namely initial, middle and final words. The division of the words into three regions, depends upon the length and other emotion related attributes of the words. The details of this division of an utterance into 3 parts are given in Table 3.5. In this study, global and local prosodic features are extracted from initial, middle and final parts of the utterance. For each portion of the utterances (initial, middle or final words), emotion analysis is carried out using global and local prosodic features in a similar manner as it was performed at the sentence level. Further the overall emotion recognition performance, from the word level prosodic features is obtained by combining the measures from the emotion recognition systems (ERSs) developed using initial, middle and final words. Figure 3.5 shows the block diagram of ERS

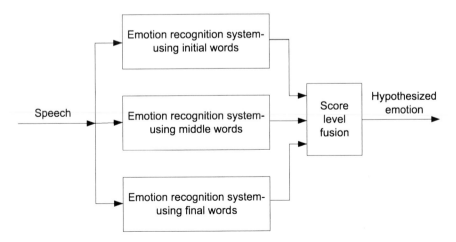

Fig. 3.5 Emotion recognition system using word level global and local prosodic features

developed using word level prosodic features. Each block of Fig. 3.5 contains the ERS shown in Fig. 3.3. From each portion of the sentence, global and local prosodic features are computed and the ERSs are developed as shown in Fig. 3.3. Later the measures from initial, middle and final words are further combined to get overall measure.

Table 3.10 shows the average emotion recognition performance of word level global and local emotion recognition systems. Table 3.11 shows the overall emotion recognition performance using word level prosodic features, obtained by score level combination of local and global features of initial, middle, and final words.

In Table 3.10, column *Global* indicates the results obtained by using only global prosodic features. Under local prosodic parameters, systems are individually developed using duration, pitch and energy components. Columns with headings *Dur.*, *Pitch*, and *Energy* show the results obtained using the dynamic nature of duration, pitch and energy parameters respectively. Column 'Local' indicates the results obtained due to the score level combination of duration, pitch, and energy parameters with appropriate weighting factors. Column *Glo. + Loc.* indicates the emotion recognition performance due to the combination of measures from global (*Glo.*) and local (*Loc.*) systems. All these results are reported for initial, middle and final words of the utterances of IITKGP-SESC.

From the results of Table 3.10, it is evident that, all parts of the utterances do not contribute uniformly toward emotion recognition. Some of the important observations are mentioned below. There is a drastic improvement in the recognition performance by using local prosodic features compared to global features. However, improvement in the recognition performance is marginal over the local features, when global and local features are combined. This indicates that, global prosodic features at word level, may not be complementary in nature, with respect to their local counterparts. In the case of individual local prosodic features, energy features are more discriminative with initial words of the utterances. This is obvious, as generally, all utterances have dominant energy profiles in the beginning. In the case of middle words, energy and pitch parameters have almost equal emotion discrimination with the recognition rate of 43 and 44 % respectively. Pitch values have the dominant distinction in the case of emotion recognition using final words. Duration information has always been least discriminative in case of initial, middle, and final words. In general final words carry more emotion discriminative information of about 64 %, compared to their initial and middle counterparts. It is observed that recognition performance using final words is almost the same as the performance achieved using the entire sentence. It indicates that only about (1/3)rd portion of the sentence (final part) is sufficient to recognize the emotions. Interestingly, the average performance obtained due to the combination of the measures of initial, middle, and final words is almost equal to the recognition rate obtained using entire utterances (See Table 3.8). Comparison of emotion recognition performance of individual emotions with respect to initial, middle, and final words is given in Fig. 3.6. It may be seen from the figure that passive emotions like disgust, sadness, neutral, and surprise are better discriminated using final words of the sentences. Initial words played an important role in

Table 3.10 Emotion classification performance using global and local prosodic features computed from the words of the utterances depending on their position

Emotions	Global	Local features			Local	Glo.+ Loc.
		Dur.	Pitch	Energy		
Initial words, average emotion recognition: 50.13						
Anger	40	20	33	47	53	57
Disgust	33	33	27	33	37	37
Fear	37	33	17	53	50	43
Happiness	20	13	40	20	40	47
Neutral	23	23	47	30	43	43
Sadness	30	23	37	33	40	47
Sarcasm	60	40	43	73	63	67
Surprise	30	23	43	30	47	60
Avg.	34	26	36	39	47	50
Middle words, average emotion recognition: 58.38						
Anger	50	17	43	53	57	60
Disgust	57	53	53	43	63	70
Fear	60	23	30	67	67	70
Happiness	33	30	50	30	40	43
Neutral	43	20	53	37	60	67
Sadness	40	27	47	37	50	57
Sarcasm	30	67	33	57	63	63
Surprise	30	40	40	23	50	37
Avg.	43	35	44	43	56	58
Final words, average emotion recognition: 64						
Anger	33	40	30	43	43	43
Disgust	53	37	83	30	80	83
Fear	67	10	47	57	60	63
Happiness	23	33	33	23	30	33
Neutral	70	33	80	53	77	77
Sadness	73	27	83	30	80	83
Sarcasm	23	47	47	60	63	60
Surprise	63	23	67	60	70	70
Avg.	51	31	59	45	63	64

Avg. Average, *Dur.* Duration, *Glo.* Global, *Loc.* Local

Table 3.11 Overall emotion classification performance by combining local and global prosodic features computed from the words from different positions

	Ang.	Dis.	Fear	Hap.	Neu.	Sad	Sar.	Sur.	Avg.
Initial + Middle + Final + words	57	77	70	47	67	80	60	63	65.38

recognizing the emotions like happiness and sarcasm. Anger and fear are recognized well using middle parts of the sentences.

3.5.3 Emotion Recognition Systems using Syllable Level Prosodic Features

Within each word, emotion specific information at the initial, middle and final syllables may be different for different emotions. Based on this intuition, we have carried out the analysis of emotion specific information at the syllable level also. While analyzing emotions at the syllable level, two types of models are developed. (1) Utterance-wise syllable-level emotion recognition models and (2) Region-wise syllable-level emotion recognition models. In the case of utterance wise syllable models, initial, middle and final syllables of all the words of the utterance are grouped separately. Then the emotion models are developed using global and local prosodic features derived from these syllable groups. In the case of region-wise syllable models, initial, middle, and final syllables taken from the specific portion of the sentence are grouped. In this manner we have 3 sets of initial, middle, and final syllables corresponding to these regions of the sentence. Here the regions indicate initial, middle, and final words. Then the emotion models are developed using the global and local features of these syllable groups.

3.5.3.1 Utterance-Wise Syllable-Level Emotion Recognition

Some of the emotions expressed using high arousal properties such as anger and happiness, cannot retain the same energy throughout the word. Therefore, in such cases high energy profiles are well exhibited in the case of initial syllables of the words. Similarly syllable-level variations may be observed in the case of duration and pitch patterns of different emotions. Hence, in this work, we analyzed the syllables for their emotion discriminative nature. For analyzing the emotions at the syllable level, syllables present in each sentence are divided into 3 groups as initial, middle and final syllables, based on their position in the word. The details of syllable groups are given in Table 3.5. The block diagram of the ERSs developed using syllable-level prosodic features is similar to the one shown in Fig. 3.5. Here ERS in each block has the same structure as shown in Fig. 3.3.

The emotion recognition performance of utterance-wise syllable-level models is given in Table 3.12. From the results, it is observed that at the syllable level also local prosodic parameters perform better emotion recognition than the global prosodic features. Initial and final syllables carry more emotion-specific information than the middle syllables. Emotion recognition performance using initial syllables is slightly better than the performance of the systems developed using only initial words. This may be due to the dominance of energy and pitch profiles in the initial portion of

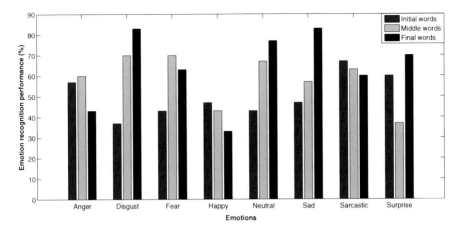

Fig. 3.6 Comparison of emotion recognition performance using global+local prosodic features extracted from *initial*, *middle* and *final* words

the words than the entire duration of words. In general, final syllables contribute heavily toward emotion recognition. Average emotion recognition due to only final syllables is about 61 %, whereas it is 51 and 46 % respectively for the groups of initial and middle syllables. The comparison of recognition of different emotions using utterance-wise initial, middle, and final syllables is given in Fig. 3.7. From the figure, it is observed that, final syllables have more emotion discriminative information, in case of most of the emotions except anger and happiness. These are high-arousal emotions, and hence, their discrimination is better in the case of initial syllables. Middle syllables do not contribute much toward emotion recognition than the initial and final syllables. The overall emotion recognition performance due to the combination of initial, middle, and final syllables is given in Table 3.13, This is comparable to the results of word and utterance level studies (See Tables 3.9 and 3.11).

3.5.3.2 Region-Wise Syllable-Level Emotion Recognition

At the word-level emotion recognition analysis, we have studied emotion discriminative characteristics from the set of initial, middle and final words. Within these groups of words, there may be some additional emotion discriminative information present at the syllable level. Therefore, to capture emotion-specific information from the syllables within the group of words (initial, middle and final), the syllables of these words are divided into initial, middle and final syllables. The syllables within each region of words (initial, middle and final words) are grouped into initial, middle, and final syllables, based on their positions in the words. The details of syllable groups are given in Table 3.5. The block diagram of the ERSs developed using region-wise syllable-level prosodic features is similar to the ERS developed using word level features. Here, the ERS in each block has the same structure as shown in Fig. 3.3. Overall

Table 3.12 Emotion classification performance using global and local prosodic features computed from syllables taken from all words of the utterances, grouped based on their position in the word

| Emotions | Global | Local Features | | | | Glo.+ |
		Dur.	Pitch	Energy	Local	Loc.
Initial syllables, average emotion recognition: 51 %						
Anger	30	23	30	57	50	47
Disgust	20	30	43	40	43	53
Fear	53	10	30	47	50	50
Happiness	17	17	47	20	43	47
Neutral	33	27	53	27	50	47
Sadness	30	27	43	40	43	47
Sarcasm	63	53	57	67	63	63
Surprise	23	33	47	17	43	50
Avg.	34	28	44	39	48	51
Middle syllables, average emotion recognition: 46 %						
Anger	13	30	27	27	30	30
Disgust	17	43	37	43	47	43
Fear	37	27	27	47	43	47
Happiness	10	17	30	23	30	30
Neutral	40	30	50	43	53	57
Sadness	37	33	33	33	37	43
Sarcasm	30	67	40	47	63	67
Surprise	47	30	47	40	43	50
Avg.	29	35	36	38	43	46
Final syllables, average emotion recognition: 61 %						
Anger	13	27	23	40	40	43
Disgust	43	47	53	57	60	60
Fear	53	23	57	63	60	63
Happiness	17	43	37	27	43	47
Neutral	67	33	67	50	70	73
Sadness	33	27	53	37	57	60
Sarcasm	33	37	47	70	73	73
Surprise	27	27	57	30	60	67
Avg.	36	33	49	47	58	61

Avg. Average, *Dur.* Duration, *Glo.* Global, *Loc.* Local

ER performance using region-wise syllable-level features is obtained by combining the measures from the ERSs developed using initial, middle, and final syllables.

The performance of ERSs developed using region wise syllable-level, global and local prosodic features is shown in Table 3.14. In this study, three sets of studies are carried out similar to the studies on utterance-wise syllable level emotion recognition. Initial, middle and final syllable models are developed using global and local prosodic features derived from the syllables of the initial words of the utterance. Similarly, syllable models are developed using global and local prosodic features derived from the syllables of middle and final words. In Table 3.14, these results are shown in separate rows. From the table, it is observed that the syllables of the final words carry

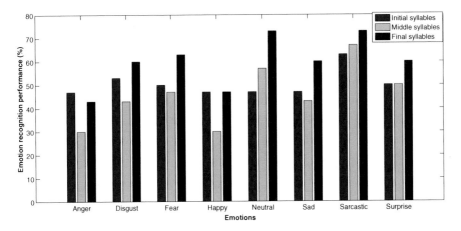

Fig. 3.7 Comparison of emotion recognition performance using utterance-wise *initial*, *middle*, and *final* syllable prosodic features

Table 3.13 Overall emotion classification performance by combining local and global prosodic features computed from the syllables from different positions

	Ang.	Dis.	Fear	Hap.	Neu.	Sad	Sar.	Sur.	Avg.
Initial + Middle + Final syllables	43	67	63	43	70	67	73	77	63

Average emotion recognition: 63.28 %

Table 3.14 Emotion classification performance using global and local prosodic features computed from syllables within specific regions like initial, middle, and final words

Wd. pos.	Global	Local Features				Glo.+ Loc.
		Dur.	Pitch	Energy	Local	
Initial words, Average emotion recognition: 26						
Ini. syls.	24	16	20	20	26	27
Mid. syls.	18	10	14	19	20	21
Fin. syls.	15	19	22	19	23	24
Middle words, average emotion recognition: 22						
Ini. syls.	14	8	19	15	20	20
Mid. syls.	17	11	22	18	23	24
Fin. syls.	12	7	14	16	18	22
Final words, average emotion recognition: 46.33						
Ini. syls.	26	18	31	31	41	48
Mid. syls.	25	16	31	29	36	38
Fin. syls.	38	27	42	40	51	53

Avg. Average, *Dur.* Duration, *Fin.Syls.* Final syllables, *Glo.* Global, *Ini.Syls.* Initial syllables, *Loc.* Local, *Mid.Syls.* Middle syllables, *Syl.Pos.* Syllable position

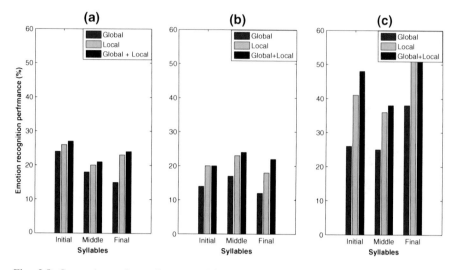

Fig. 3.8 Comparison of emotion recognition performance of proposed global, local, and global+local prosodic features with respect to syllables within the groups of initial, middle, and final words of IITKGP-SESC. **a** *Initial, middle* and *final* syllables of *initial* words, **b** *initial, middle* and *final* syllables of *middle* words, and **c** *initial, middle* and *final* syllables of *final* words

comparatively more emotion specific information, with the recognition rate of around 46 %. Within these final words, final syllables followed by initial syllables contribute more toward emotion recognition, with an accuracy of 50 and 48 % respectively. Figure 3.8 shows the comparison of emotion recognition performance using region-wise syllable-level prosodic features.

It is observed from the results that the recognition performance is very poor with the region-wise syllable level prosodic features. The basic reason for this poor performance is that use of shorter speech segments for feature extraction while training and testing the emotion models.

3.6 Summary

In this Chapter, prosodic analysis of speech signals has been performed at different levels of speech segments for the task of recognizing the underlying emotions. Eight emotions of IITKGP-SESC are used in this study. Support vector machines are used for developing the emotion models. Global and local prosodic features are separately extracted from utterance, word and syllable segments of speech for developing the emotion models. Word and syllable boundaries are identified using VOPs. Global prosodic features are derived by computing statistical parameters like mean, maximum, minimum from the sequence of prosodic parameters. Local prosodic parameters are obtained from the sequence of syllable durations, frame level pitch and energy values. The prosodic contour trends are retained through local prosodic features. The contribution of different parts of the utterances toward emotion recognition is studied by developing emotion recognition models using the prosodic features obtained from initial, middle, and final regions of the utterances. The combination of local and global prosodic features was found to marginally improve the performance compared to the performance of the systems developed using only local features. From the word and syllable level prosodic analysis, the unique observation in the view of discriminating the emotions is that final words and syllables contain more emotion discriminative information than the other groups of words and syllables.

References

1. K. Murty, B. Yegnanarayana, Epoch extraction from speech signals. IEEE Trans. Audio Speech Lang. Process. **16**, 1602–1613 (2008)
2. S.R.M. Prasanna, B.V.S. Reddy, P. Krishnamoorthy, Vowel onset point detection using source, spectral peaks, and modulation spectrum energies. IEEE Trans Audio Speech Lang Process **17**, 556–565 (May 2009)
3. S.R.M. Prasanna, J.M. Zachariah, Detection of vowel onset point in speech, in *Proceedings of IEEE International Conference Acoustics Speech, Signal Processing* Orlando, Florida, USA May 2002
4. S.R.M. Prasanna, *Event-Based Analysis of Speech*. PhD thesis, Department of Computer Science and Engineering, Indian Institute of Technology Madras, Chennai, India, Mar 2004
5. http://web.nickshanks.com/fonts/telugu/
6. K.S. Rao, S.G. Koolagudi, R.R. Vempada, Emotion recognition from speech using global and local prosodic features. Int. J. Speech Technol. **15**, 265–289 (2012). doi:10.1007/s10772-012-9172-2

Chapter 4
Robust Emotion Recognition using Combination of Excitation Source, Spectral and Prosodic Features

Abstract Different speech features may offer emotion specific information in different ways. This chapter explores the combination evidences offered by various speech features. In this chapter, we consider excitation source, spectral and prosodic features as specific individual speech features for classifying the emotions. Various combinations of the above mentioned individual features are explored for improving the emotion recognition performance. Since, the features are derived from different levels, the emotion specific characteristics captured by these features may be complementary or non-overlapping in nature. By properly exploiting these evidences, the recognition performance will definitely improved. From the results, its is observed that all the combinations explored in this have enhanced the recognition performance significantly.

4.1 Introduction

Chapters 2 and 3 have discussed robust emotion recognition using system and prosodic features independently. From the speech production and feature extraction points of view, source, system and prosodic features are completely different. System features represent the response of the vocal tract system derived from the autocorrelation analysis, which mostly capture the first and second order correlations from speech signals. On the other hand, excitation source features are represented by a sequence of LP residual samples, which represents higher order correlations. Prosodic features represent duration, intonation and intensity patterns of the sequence of speech segments like sentences, words, and syllables. Therefore, in view of feature extraction, speech production and perception aspects, the source, system and prosodic features are non-overlapping. In this regard, the emotion specific information carried by these features may also be non-overlapping in nature. With this idea, in this chapter, we have explored the combination of different speech features to study the effect on emotion recognition. This chapter discusses the combination of measures from source, system, and prosodic features taken pair-wise and all together. For this

K. S. Rao and S. G. Koolagudi, *Robust Emotion Recognition using Spectral and Prosodic Features*, SpringerBriefs in Speech Technology, DOI: 10.1007/978-1-4614-6360-3_4, © The Author(s) 2013

study, Set3 of IITKGP-SESC is used as the dataset, where 8 speakers' speech data is used for training the emotion recognition models and 2 speakers' speech data is used for validating them. Chapter 5 discusses about emotion recognition using speaking rate features. In this chapter, emotion recognition is carried out in two stages. Set3 of IITKGP-SESC is used to evaluate the performance of two-stage emotion recognition systems developed based on a speaking rate features. A linear combination of the measures from different features is used while combining the features. Chapter 6 explores the combination of speech features, for recognizing the real-life emotions. Five emotions collected from Hindi commercial and art movies are used to represent real-life emotions. Multi-speaker emotion speech data of around 12 min collected from Hindi movies is used for building each of the emotion models. The test utterances of duration 2–3 s, derived from the remaining 3 min of data are used for evaluating the trained models.

4.2 Feature Combination: A Study

From the production perspective, speech is a convolved outcome of excitation source and the vocal tract system response. Prosodic features are extracted from the longer speech segments to represent perceptual quality of the speech, such as melody, timbre, rhythm and so on. Ideally, the three speech features used in this book (excitation source, vocal tract system and prosodic) for emotion recognition represent three different aspects of speech production and perception. Therefore, it is believed that they contain non-overlapping and supplementary emotion-related information. In this chapter, we intend to exploit the supplementary nature of these features, by combining their measures to improve the emotion recognition performance [1].

Scientifically, emotions are studied from different viewpoints. Psychologists tried to map all the emotions onto three dimensional space, known as emotional space. These dimensions are arousal (activation), pleasure (valence), and dominance (power). Generally it is known that the group of emotions like anger, happiness, and fear has high arousal characteristics, similarly, disgust and sarcasm have negative arousal characteristics. For discriminating these emotions, within the group, the other dimensions such as valence may be needed.

Generally, the arousal characteristic of the emotions is represented by the prosodic parameter *intensity*. The intensity characteristics in turn influence the other prosodic parameters. For instance, the higher arousal characteristics indicate high energy, leading to high pitch and low duration. For example, high arousal emotions like anger and fear have high energy, intonation and smaller duration. These emotions cannot be discriminated using only prosodic features. Along with prosodic features, the features representing other emotion dimensions such as valence are essential. Thus, mis-classification of emotions within the set of emotions sharing similar acoustical properties, may be reduced. Table 4.1 highlights some results from the literature, where anger and happiness are mis-classified among themselves, when prosodic features are used. From the results presented in the Chap. 2, it may be noted that

Table 4.1 Percentage of mis-classification between *anger* and *happiness* emotions, quoted by different research works, in the literature, by using prosodic features

Reference	Language	Percentage of anger utterances classified as happiness	Percentage of happiness utterances classified as anger
Serdar Yeldirim, et.al. [2]	English	42	31
Dimitrios Virviridis, et.al. [3]	Scandinavian language (Danish)	20	14
Felix Burkhardt, et.al. [4]	German (Berlin Emotion Database)	–	12
Oudeyer Pierre, et.al. [5]	Concatenated synthesis (English)	35	30
S G. koolagudi, et. al. [6]	Telugu	–	34
S. G. Koolagudi	Berlin, German	27	20
Raquel Tato [7]	German	24	25

this type of mis-classification of emotions mentioned above can be reduced by some extent by using spectral features. Hence, it is hypothesized that the combination of different features may improve the emotion recognition performance and make the systems more robust.

In this work, speech features extracted from excitation source, VT system, and prosodic aspects are combined for analyzing the emotion recognition. Among various excitation source, spectral and prosodic features proposed in the previous chapters, the following features are considered in this chapter for combining measures. In each case, the best performing features are chosen for combination. They are (a) LP residual samples chosen around glottal closure instants, (b) Twenty one LPCCs extracted from the whole speech utterance, using a block processing approach, and (c) local prosodic features, capturing duration, pitch and energy profiles of the speech utterance. While combining the features, score level fusion is preferred as the features are extracted using different mechanisms. For example, spectral features are extracted from the frame level, excitation source features are extracted at the epoch level, and prosodic features are extracted from the utterance level. Hence, feature level fusion is not suitable to combine the features derived using heterogeneous approaches.

In this work, AANNs are used to capture the emotion specific information from excitation source features, GMMs are used for developing the models using spectral features, and SVMs are used to discriminate the emotions using prosodic features. Since the measures are derived from different models and features, they need to be normalized appropriately before combining. The weighted combination of scores for (a) excitation source and vocal tract system features, (b) source and prosodic features, (c) system and prosodic features, and (d) source, system and prosodic features is studied and the results are discussed in the following sections.

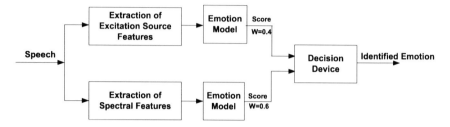

Fig. 4.1 Emotion recognition system using the combination of source and spectral features

4.3 Emotion Recognition using Combination of Excitation Source and Vocal Tract System Features

From the studies carried out in previous chapters, it is evident that excitation source and vocal tract components of speech have certain emotion discriminative power. These two features represent two different aspects of speech production mechanism. Therefore, to exploit the unique emotion specific information provided by these two features, their measures are combined through appropriate weighting factors. The process of combination of measures is shown in Fig. 4.1. The weighting rule for combining the confidence scores (measure) of individual features is as follows: $c_f = \sum_{i=1}^{m} w_i c_i$, where c_f is the combined confidence score derived from the confidence scores of the individual features, w_i and c_i are weighting factor and confidence score corresponding to the ith feature respectively, and m indicates the number of individual features used for combining the scores. In this work, we have combined the measures of the ER systems developed by two features, one of the weights (w_i) is varied from 0 to 1, in steps of 0.1, and the other weights are determined using the formula: $w_j = \frac{1-w_i}{m-1}$, where $j = 2$ and $i = 1$. The emotion recognition performance obtained using different combinations of weighting factors is given in Fig. 4.2. It is observed that, the best recognition performance is about 74 % for the weighting factors of 0.6 and 0.4 to the confidence scores of system and source features respectively. From the results (see Table 4.2), it is observed that recognition performance of anger, disgust, fear, happiness, and sadness has improved in the combined system, whereas for neutral, sarcasm and surprise there is not considerable improvement. The recognition performance of the combined system is observed to be increased by around 4 % compared to the system developed using spectral features alone. The improvement in the performance may be due to the supplementary nature of the measures provided by the excitation source features. The details of the classification performance, using the fusion of spectral and excitation source information are given in Table 4.2. From the table it may be noted that spectral features are more discriminative with respect to emotions compared to excitation source features. Spectral and excitation source features have performed similarly in the case of anger and fear. Comparison of emotion recognition performance obtained using spectral, excitation source, and spectral+source features may be visualized from the bar graph shown in Fig. 4.3.

Table 4.2 Emotion classification performance using the combination of spectral and excitation source features

Emotions	Emotion recognition performance (%)		
	System features	Excitation source features	Comb.(Spec. + Excit.) (0.6 + 0.4)
Ang.	57	57	70
Dis.	67	54	70
Fear	63	60	73
Happy	73	43	80
Neut.	83	63	83
Sad	73	47	77
Sarcastic	77	50	77
Surprise	63	57	63
Avg.	69.5	53.88	73.75

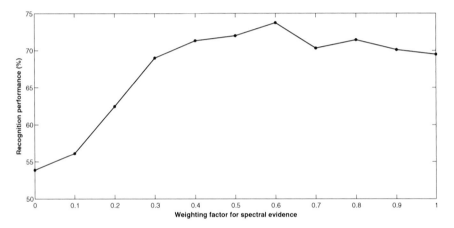

Fig. 4.2 Variation in emotion recognition performance with different weighting factors, for the combination of spectral and excitation source features

4.4 Emotion Recognition using Combination of Vocal Tract System and Prosodic Features

In this work, local prosodic and frame level spectral features are separately extracted from the utterances. Frame-wise pitch and energy values, syllable wise durations are extracted to represent prosodic information. Similarly, 21 LPCCs are used as the correlates of spectral information. Emotion recognition models are separately developed using prosodic and spectral features. The measures are combined with

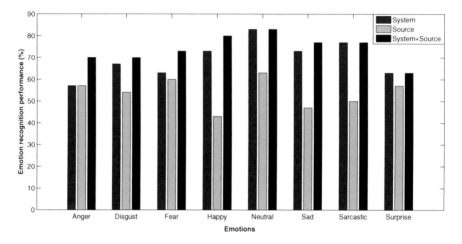

Fig. 4.3 Comparison of emotion recognition performance with respect to each emotion using excitation source, spectral, and source + spectral features derived on IITKGP-SESC

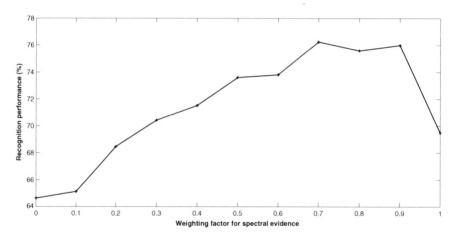

Fig. 4.4 Variation in emotion recognition performance with different weighting factors, for the combination of spectral and prosodic features

appropriate weights to obtain the emotion recognition performance of the combination of features.

The emotion recognition performance obtained using different combinations of weighting factors is shown in Fig. 4.4. The weights of 0.7 and 0.3 are found to achieve better emotion recognition for spectral and prosodic measures respectively, with score level combination. Table 4.3 shows the consolidation of emotion recognition results, obtained using prosodic and spectral features individually, and also in combination.

Table 4.3 Emotion classification performance using the combination of prosodic and system features

Emotions	Emotion recognition performance (%)		
	System features	Prosodic features	Syst. + Pros. features (0.7 + 0.3)
Anger	57	67	70
Disgust	67	73	70
Fear	63	70	77
Happiness	73	77	83
Neutral	83	60	80
Sadness	73	60	83
Sarcasm	77	57	77
Surprise	63	53	70
Average	69.5	64.63	76.25

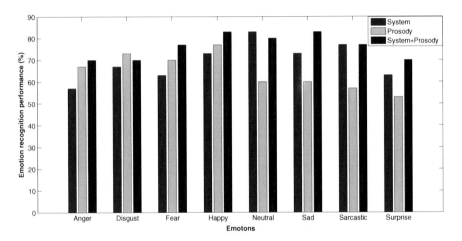

Fig. 4.5 Comparison of emotion recognition performance with respect to each emotion using spectral, prosodic, and spectral + prosodic features derived on IITKGP-SESC

In general, spectral features outperformed prosodic features while classifying the emotions. However, in the case of active emotions like anger, disgust, fear, and happiness, spectral features performed poorly compared to their prosodic counterparts. It indicates that the prosodic features dominate the spectral features during classification of the above mentioned emotions. For slow emotions, especially like sadness and sarcasm, spectral features performed quite well. In the case of neutral and sarcasm considerable improvement in the recognition performance is not observed even after the combination of features. These observations are shown in the form of a bar graph in Fig. 4.5. After combining spectral and prosodic features, there is a considerable improvement of more than 6% compared to the performance of spectral features. This indicates that prosodic features provide considerable supplementary emotion specific measure to the spectral features.

4.5 Emotion Recognition using Combination of Excitation Source and Prosodic Features

In this work, excitation source and prosodic features are separately extracted from utterances and used to develop emotion recognition systems. LP residual samples around glottal closure instants are used as the correlates of excitation source information. On the other hand, frame-wise pitch and energy values, sequence of duration of syllables represent prosodic information. The details of extraction and use of excitation source features are given in [8]. Extraction of prosodic features is discussed in Sect. 3.5. The measures are combined with suitable weights to obtain the emotion recognition performance of combinations of features.

Variation in emotion recognition performance with different weighting factors, for the combination of prosodic and excitation source features is given in Fig. 4.6. The weights of 0.6 and 0.4 are found to achieve better emotion recognition for prosodic and excitation source feature measures respectively, during score level combination. Table 4.4 shows the emotion recognition results, obtained using prosodic and excitation source features individually and also in combination. Emotion recognition results presented in Table 4.4 indicate that, in general, prosodic features perform well compared to excitation source features. However, in the case of surprise and neutral, the performance of source features is slightly better. The combination of measures from prosodic and source features has improved the recognition performance by 4 %, compared to the performance by the prosodic features. This is the indication of non-overlapping characteristics of excitation and prosodic features with respect to the discrimination of emotions. These trends may be observed from the bar graph given in Fig. 4.7.

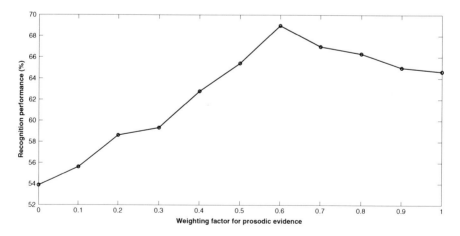

Fig. 4.6 Variation in emotion recognition performance with different weighting factors, for the combination of prosodic and excitation source features

Table 4.4 Emotion classification performance using the combination of prosodic and source features

Emotions	Emotion recognition performance (%)		
	Source features	Prosodic features	Sour. + Pros. features (0.4 + 0.6)
Anger	57	67	70
Disgust	54	73	73
Fear	60	70	73
Happiness	43	77	80
Neutral	63	60	70
Sadness	47	60	63
Sarcasm	50	57	63
Surprise	57	53	60
Average	53.88	64.63	69

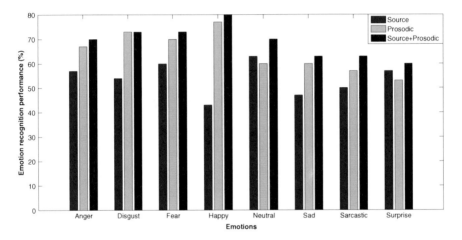

Fig. 4.7 Comparison of emotion recognition performance with respect to each emotion using excitation source, prosodic, and source + prosodic features derived on IITKGP-SESC

4.6 Emotion Recognition using Combination of Excitation Source, Vocal Tract System and Prosodic Features

In this work, excitation source, VT system, and prosodic features are separately extracted from utterances and used to develop emotion recognition systems. The details of extraction and use of excitation source features are given in [8]. The details of extraction of system and prosodic features are given in Sects. 2.5 and 3.5, respectively. The measures obtained from the emotion recognition systems developed using individual features are combined with suitable weights to evaluate the emotion recognition performance of the combined features.

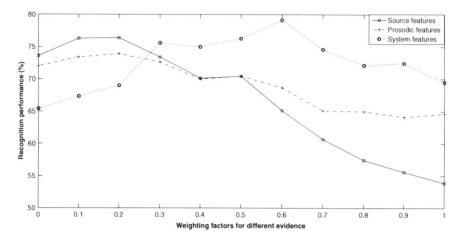

Fig. 4.8 Variation in emotion recognition performance with different weighting factors, for the combination of spectral, excitation source, and prosodic features. Other two features than mentioned in the legend are equally weighted

Table 4.5 Emotion classification performance using the combination of source, system, and prosodic features on IITKGP-SESC

Emotions	Emotion recognition performance (%)			
	Source features	System features	Prosodic features	Sour. + Sys. + Pros. features (0.2 + 0.6 + 0.2)
Anger	57	57	67	73
Disgust	54	67	73	70
Fear	60	63	70	77
Happiness	43	73	77	83
Neutral	63	83	60	87
Sadness	47	73	60	83
Sarcasm	50	77	57	87
Surprise	57	63	53	73
Average	53.88	69.5	64.63	79.13

The emotion recognition performance obtained using different combinations of weighting factors is given in Fig. 4.8. The weights of 0.6, 0.2, and 0.2 are found to achieve the best emotion recognition performance for spectral, excitation source, and prosodic measures respectively, during score level combination. Table 4.5 shows the emotion recognition results, obtained using source, spectral, and prosodic features individually and also in combination. It is evident from the Table that the score level combination of three different features works well and there is an average improvement of around 10% in the emotion recognition performance compared to the performance of spectral features alone. The fast emotions like anger, disgust, fear, and happiness are recognized well by prosodic features and the slow emotions

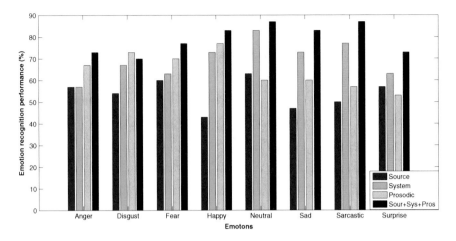

Fig. 4.9 Comparison of emotion recognition performance with respect to each emotion using excitation source, spectral, prosodic, and source + spectral + prosodic features derived on IITKGP-SESC

like neutral, sadness and sarcasm are recognized well by spectral features. Surprise is a exception for the above trend. The reason for good recognition performance of spectral features for slow emotions may be due to the availability of sufficient time for the vocal tract in taking specific shape while expressing the emotions. It is quite interesting to know that, after combination of the features, the performance is better than the best among the three individuals for all emotions, except for disgust. It indicates that all three components of speech provide supplementary contribution toward emotion recognition. Visualization of the above discussion of results may be realized through the bar graph given in Fig. 4.9.

The Berlin emotion speech corpus is widely used by many researchers for comparing their proposed features and methods used for classifying the emotions. In this study, the combination of excitation source, vocal tract (VT) system, and local prosodic features is used to study the emotion classification on the Berlin emotion database (Emo-DB). Out of 10 speakers, the speech data of 8 speakers is used for training the models, the remaining 2 speakers' speech data is used for testing the developed models. The emotion recognition performance of individual and combination of features is given in Table 4.6. From the results of Table 4.6, it is observed that there is 15 % of improvement in the emotion recognition performance by combining the source, system and prosodic features derived from the Berlin emotional speech corpus (Emo-DB). From the results of Tables 4.5 and 4.6, it is observed that the ER performance using combinations of features derived from IITKGP-SESC and Emo-DB is almost same. The bar graph given in Fig. 4.10 compares the emotion-wise recognition performance of the emotions present in Emo-DB using source, system, prosodic, and source + system + prosodic features.

Table 4.6 Emotion classification performance using the combination of excitation source, spectral and prosodic features on Emo-DB

Emotions	Emotion recognition performance (%)			
	Excitation source	Spectral	Prosodic	Comb. (Spec. + Excit. + Pro.) (0.6 + 0.2 + 0.2)
Anger	43	67	53	90
Boredom	53	63	63	80
Disgust	60	64	70	67
Fear	47	70	57	83
Happiness	60	53	70	87
Neutral	57	73	67	70
Sadness	47	57	57	77
Avg.	52.43	63.85	62.43	79.14

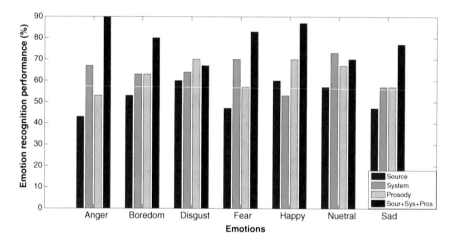

Fig. 4.10 Comparison of emotion recognition performance with respect to each emotion using excitation source, spectral, prosodic, and source + spectral + prosodic features derived on Emo-DB

Table 4.7 shows the emotion classification performance on Emo-DB by some recent studies. From the results presented in the Table, it is observed that, the emotion classification performance is reported between 50–81 % using various spectral and prosodic features, with different linear and nonlinear models. The last row of Table 4.7 indicates the performance of the proposed ERSs developed using the combination of measures from Excitation source (ES), VT and local prosodic features. Combining the measures from the ES, VT, and local prosodic features yields the emotion recognition performance of around 79 %, in case of the Emo-DB, which is comparable to the best results quoted in Table 4.7 (See row 2 of the Table).

Table 4.7 The comparison of emotion classification performance using different features and models on Berlin emotional speech corpus

Features	Models	Average emotion recognition performance	References
Pitch, energy and temporal emotion descriptors	Binary decision tree	72.04	[9]
Pitch, energy and MFCC features	Bayes classifier	81.14	[10]
Formants and prosodic features	Multi-layer perceptron	80.67	[11]
MFCCs	HMM	55.14	
	SVM	68.44	[12]
$MFCCs + \Delta + \Delta^2$	HMM	59.26	
	SVM	73.77	
Pitch, LPCs and long term modulation spectrum	Artificial Neural Networks	50.00	[13]
Prosodic (pitch, intensity, jitter, shimmer) features, and formant energies	TGI+.1 (weka)	78.58	[11]
	C4.5	52.9	
	Multi-layer perceptron	73.9	
Acoustic features, (22 MFCCs) prosodic (pitch, intensity, rhythm) features from pseudo phonetic regions	k-nearest neighbor classifier	73.80	
		55.80	[14]
Combination of spectral & Excitation source features	AANN	79.14	Present work

4.7 Summary

Combination of measures from different speech features is well addressed in the literature, to improve the performance of different speech systems. In this chapter, through analysis of a combination of ES, VT system and prosodic features is carried out, with the intention of studying the effect of combining the features on emotion recognition performance. As the features are extracted using heterogeneous approaches, a weighted combination of scores is used as the fusion method. ES, VT system, and prosodic features have shown individual supplementary properties, when combined with other features. Even after combination, VT system features seem to be dominant over other two features. After combining the features, emotion recognition performance has been improved by an average of 4–6 %, compared to the nearest highest individual performance.

References

1. S.G. Koolagudi, K.S. Rao, Emotion recognition from speech using source, system and prosodic features. Int. J. Speech Technol. **15**(3), 265–289 (Springer, 2012)
2. S. Yildirim, M. Bulut, C.M. Lee, A. Kazemzadeh, C. Busso, Z. Deng., S. Lee, S. Narayanan, An acoustic study of emotions expressed in speech. International conference on spoken language processing (ICSLP 2004), Jeju island, Korean, Oct 2004
3. D. Ververidis, C. Kotropoulos, I. Pitas, *Automatic emotional speech classification* (ICASSP, IEEE, 2004), pp. 1593–1596
4. F. Burkhardt, W.F. Sendlmeier, Verification of acousical correlates of emotional speech using formant-synthesis. ITRW on Speech and Emotion, Newcastle, Northern Ireland, UK, 5–7 Sept 2000, pp. 151–156
5. P.-Y. Oudeyer, The production and recognition of emotions in speech: features and algorithms. Int. J. Hum Comput Stud. **59**, 157–183 (2003)
6. S.G. Koolagudi, S. Maity, V.A. Kumar, S. Chakrabarti, K.S. Rao, in *IITKGP-SESC: Speech Database for Emotion Analysis*. Communications in Computer and Information Science, JIIT University, Noida, 17–19 Aug 2009. Springer. ISSN: 1865–0929
7. R. Tato, R. Santos, R. Kompe1, J. Pardo, Emotional space improves emotion recognition. 7th International conference on spoken language processing, Denver, Colorado, USA, 16–20 Sept 2002
8. K. S. Rao, S. G. Koolagudi, Characterization and recognition of emotions from speech using excitation source information. Int. J. Speech Technol. **15**, (Springer, Sept 2012) doi:10.1007/s10772-012-9175-2
9. K.S. Jarosaw Cichosz, Emotion recognition in speech signal using emotion-extracting binary decision trees, in *Affective Computing and Intelligent Interfaces ACII*, Lisbon, Doctoral Consortium, 13–14 Sept 2007
10. T. Vogt, E. Andre, Improving automatic emotion recognition from speech via gender differentiation, in *Proceedings of Language Resources and Evaluation Conference (LREC)*, 24–26 May 2006
11. J. Sidorova, Speech Emotion Recognition. PhD thesis, Universitat Pompeu Fabra, 4 July 2004
12. B. Schuller, S. Reiter, G. Rigoll, Evolutionary feature generation in speech emotion recognition, in *IEEE International Conference on Multimedia and Expo, (Toronto Ont)*, pp. 5–8, IEEE, 9–12 July 2006. doi:10.1109/ICME.2006.262500
13. S. Scherer, H. Hofmann, M. Lampmann, M. Pfeil, S. Rhinow, F. Schwenker, G. Palm, Emotion recognition from speech:stress experiment, in *Proceedings of the Sixth International Language Resources and Evaluation (LREC'08)* (N. C. C. Chair), ed. by K. Choukri, B. Maegaard, J. Mariani, J. Odjik, S. Piperidis, D. Tapias. (Marrakech, Morocco), European Language Resources Association (ELRA), 28–30 May 2008. ISBN: 2-9517408-4-0, http://www.lrec-conf.org/proceedings/lrec2008/
14. F. Ringeval, M. Chetouani, A vowel based approach for acted emotion recognition, in *Verbal and Nonverbal Features of Human–Human and Human–Machine Interaction: COST Action 2102 International Conference*, (Springer, Heidelberg, 2008), pp. 243–254

Chapter 5
Robust Emotion Recognition using Speaking Rate Features

Abstract In this chapter speaking rate characteristics of speech are explored for discriminating the emotions. In real life, we observe that certain emotions are very active with high speaking rate and some are passive with low speaking rate. With this motivation, in this chapter, we have proposed a two stage emotion recognition system, where the emotions are classified into three broad groups (active, neutral and passive) at the first stage and during second stage emotions in each broad group are further classified. Spectral and prosodic features are explored in each stage for discriminating the emotions. Combination of spectral and prosodic features is observed to be performed better.

5.1 Introduction

While classifying the emotions, acoustically similar emotions lead to mis-classification. This mis-classification may be reduced by performing early classification of these highly confusing emotions into different groups, before final individual classification. With this motivation the multi-stage emotion classification system is proposed where acoustically overlapping emotions are grouped into separate categories in the first stage and later individual emotion classification is performed. In this study, *speaking rate* has been chosen as the criterion for the grouping of acoustically similar emotions in the initial stage. Spectral and prosodic features, along with their combination have been used for the classification [1]. Here, spectral features are represented by 21 LPCCs and prosodic features are represented by local prosodic parameters. Set3 of IITKGP-SESC has been used as the dataset to evaluate the proposed features and approach of two-stage classification. Set3 represents text and speaker independent emotion speech data. In the first stage, emotions are categorized into 3 broad groups namely, active, normal, and passive emotions using spectral and prosodic features. These three broad groups are defined based on speaking rate. Later in the second stage, individual emotion classification is performed within each broad

K. S. Rao and S. G. Koolagudi, *Robust Emotion Recognition using Spectral and Prosodic Features*, SpringerBriefs in Speech Technology,
DOI: 10.1007/978-1-4614-6360-3_5, © The Author(s) 2013

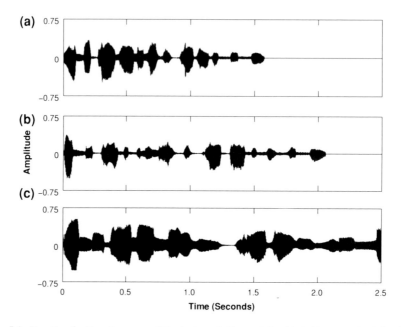

Fig. 5.1 Duration for the utterances of the text *anni dAnamulalo vidyA dAnamu minnA* for three different emotions **a** disgust, **b** neutral and **c** sarcasm

group. Excitation source features are not used in this work because, from the work discussed in the previous chapters, it may be observed that there is no appreciable influence of excitation source features on emotion recognition, during combination.

Speaking rate is a measure of number of syllables or words spoken per unit time. It is one of the important speaker specific characteristics [2]. Speaking rate variability is clearly observed while expressing different emotions (See Fig. 5.1). It is highly influenced by parameters such as duration and frequency of pauses, vowel durations, gender and age of the speaker [3, 4]. Speaking rate also depends upon physiological and psychological characteristics of the speaker [5]. Generally, younger people have faster speaking rate and older people speak more slowly for the above said reasons. Male speakers' speaking rate is slightly faster than that of female speakers [5]. By conscious manipulation of source and system parameters, speaking rate can be varied by insertion or deletion of pauses of varying lengths at different levels [6, 7]. This may be achieved due to variations in articulator movements and change in the excitation source characteristics [6]. The effect of changing speaking rate is clearly observed on prosodic features. Pitch values of spontaneous speech are highly influenced by the speaking rate variation [8]. Fast speech is characterized by an overall reduction in the pitch range [7]. Perceptual observation of fast speech indicates an increase in the intensity compared to normal speech [9]. It is also reported that dynamic behavior of the feature vectors is greatly disturbed by a change in the speaking rate [10]. Performance of speech systems is also shown to be affected by a change in speaking rate. Matthew Richardson et al. have reported that the accuracy of speech recognition depends mainly on speaking rate in addition to background noise, mismatch in microphone, language models and variations in speaker accent [11].

From careful observation of emotional speech conversation, it may be noted that speakers involuntarily change the speaking rate while expressing different emotions, compared to their normal style of speaking. For instance an angry person speaks faster with louder voice, whereas the sadness emotion is characterized by slower speaking rate including frequent pauses compared to normal speech. Therefore, it may be hypothesized that involuntary manifestation of speaking rate happens during expression of emotions and this may be used to classify the emotions at a broader level such as fast, normal, and slow emotions.

In general spectral features are treated as strong correlates of varying shapes of the vocal tract and the rate of change in the articulator movements. Prosodic parameters like intensity, duration and pitch help to capture the pronunciation pattern during variations in speaking rate. Performing emotion classification at a single step leads to low recognition performance due to poor discrimination of feature vectors among different emotions. While designing the two-stage emotion classifier, care has been taken that, misclassification rate in the first stage is as low as possible. For instance anger and happiness exhibit similar arousal characteristics, hence their discrimination using prosodic features is not effective [9]. During two-stage classification, anger and happiness are classified into two different broad categories in the first stage, to avoid their mis-classification later in the second stage. Therefore, the proposed two-stage approach improves the overall emotion classification performance, than its single-stage counterpart.

Rest of the chapter is organized as follows: Sect. 5.2 provides the motivation for the proposed two stage emotion recognition study using speaking rate features. Proposed two stage emotion recognition is discussed in Sect. 5.3. Gross level emotion recognition during first stage is discussed in Sect. 5.4. In Sect. 5.5, finer level classification of emotions is discussed. The contents of the chapter is summarized in Sect. 5.6.

5.2 Motivation

From the literature, it is observed that speaking rate depends largely upon speaker and gender. The speech waveforms shown in Fig. 5.1 correspond to the text *Anni dhanamulalo vidya daanamu minna* taken from a simulated Telugu emotion database, IITKGP-SESC. From the figure, it is observed that though the same text is uttered by the same person, the durations are different, only because of different embedded emotions. Speakers took the least time for expressing disgust and the duration is maximum for sarcasm. This observation indicates that expression of emotions directly influences the speaking rate. Therefore properties of speaking rate may be explored to characterize emotions. Table 5.1 shows the average duration of the utterances for each of the emotions in the database IITKGP-SESC. Change in duration is due to

Table 5.1 Average duration of the speech utterances of IITKGP-SESC for different emotions

Emotion	Duration (Sec)
Anger	1.80
Compassion	2.13
Disgust	1.67
Fear	1.89
Happiness	2.09
Neutral	2.04
Sarcasm	2.20
Surprise	2.09

variation in speaking rate. Hence the speaking rate may be an important parameter when classifying the emotions.

The rate of vocal folds' vibration influences the behavior of the vocal tract when producing sound units. The phenomenon of opening of vocal folds is observed to be generally the same due to their tensile/muscular restriction. But it is observed from the literature that there is a considerable variation during the closure of vocal folds. Figure 2.1 of Chap. 2 shows the spectra obtained from the steady portion of the syllable segment expressed in eight emotions. It may be observed from the figure that higher order formants (F_2, F_3 and F_4) are found to be more distinctive with respect to emotions. For analyzing the classification of speech utterances based on speaking rate, a Hindi speech database with varying speaking rates is collected at IIT Kharagpur, and is named as IITKGP-SRSC (IITKGP-Speaking Rate Speech Corpus). The database contains speech utterances of 5 different speaking rates. They are super-slow, slow, normal, fast, super-fast. Ten graduate students of IIT Kharagpur have contributed to the recording of 10 Hindi sentences in 5 different speaking rates. So in total IITKGP-SRSC contains 500 ($10\ speakers \times 10\ sentences \times 5\ speaking\ rates$) utterances. Formant analysis of slow and fast utterances is performed on IITKGP-SRSC. Figure 5.2 shows the distribution of the first and second formant frequencies for slow and fast utterances. Observation of these histograms indicates distinctive properties of formant frequencies with respect to speaking rate.

For analyzing the influence of speaking rate on spectral features, a classification system is developed based on speaking rate. LPCCs are extracted from the speech utterances of IITKGP-SRSC, using the frame size of 20 ms with the shift of 5 ms. Gaussian mixture models are used to develop the models, where the classification of speech utterances is performed based on speaking rate using spectral features. Table 5.2 shows the classification performance. Average classification performance is found to be about 82 %. The basic observation from Figs. 5.1 and 5.2 and Tables 5.1 and 5.2 motivated us to explore spectral and prosodic features for recognizing speech emotions based on speaking rate.

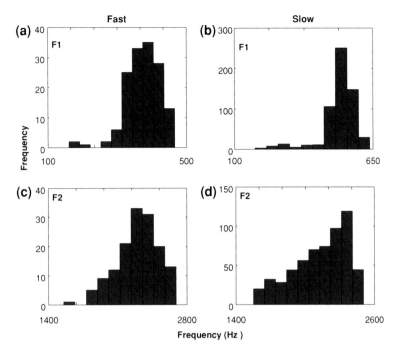

Fig. 5.2 Distribution of frame wise F_1 and F_2 values for fast and slow utterances; **a** F_1 for fast, **b** F_1 for slow **c** F_2 for fast **d** F_2 for slow Text: *mAtA aur pitA kA Adar karnA chAhiye*

Table 5.2 Classification of speech utterances based on the speaking rate using spectral features

Speaking rate	Recognition performance (%)				
	Super slow	Slow	Normal	Fast	Super fast
Super-slow	53	33	14	0	0
Slow	20	63	10	07	00
Normal	00	00	97	03	00
Fast	00	00	03	97	00
Super-fast	00	00	00	00	100

Average emotion recognition: 82 %

5.3 Two Stage Emotion Recognition System

In this work, speech emotion recognition is carried out in two stages. In the first stage, the emotions are grouped into three broad groups, namely, active, normal and passive corresponding to fast, normal, and slow speaking rates. Generally, active emotions are enthusiastically expressed with more energy, whereas passive emotions are expressed with dull mood especially with less intensity. Categorizing emotions in these three broad groups is known as gross level emotion classification. In the second

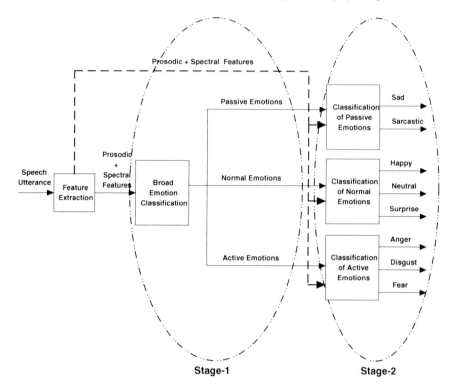

Fig. 5.3 Block diagram of two stage emotion recognition system using speaking rate

stage, individual emotion classification is performed within each broad group. This two stage classification approach basically helps to avoid mis-classification of feature vectors, which may happen in case of direct (i.e., single stage) classification.

The block diagram shown in Fig. 5.3 represents the overall emotion classification system based on the proposed two stage approach. At the first stage, depending on the speaking rate features, the unknown utterance is classified into one of the three categories, namely active (fast), normal or passive (slow) emotions. The second level of classification is known as finer level classification. From each of the broad categories utterances are further classified into individual emotion categories.

5.4 Gross Level Emotion Recognition

Based on duration analysis of emotions shown in Table 5.1, eight emotions of IITKGP-SESC are broadly categorized into three groups namely active (fast), normal and passive (slow) emotions. Anger, disgust and fear have faster speaking rate. Therefore, they are categorized as active emotions. Similarly happiness, neutral and surprise are treated as normal emotions. The remaining two emotions,

Table 5.3 Gross-level emotion classification performance using the combination of prosodic and system features

Emotion category	Emotion recognition performance (%)		
	Spectral features	Prosodic features	Spect.+Pros. features
Active	93	90	100
Normal	97	87	98
Passive	84	83	99
Average	91.30	86.67	99.00

namely sadness and sarcasm are grouped as passive emotions. In this case, classification is performed into one of the emotion groups rather than into individual emotions, therefore it is known as gross level emotion categorization. Gross level emotion classification is studied separately, using three feature sets: spectral, prosodic and their combination. Three GMMs are trained to capture the characteristics of 3 emotional categories. While developing emotion recognition models, 2400 (10 *sentences* × 10 *sessions* × 3 *emotions* × 8 *speakers*) utterances are used to train *active* and *normal* emotion models, and 1600 (10 *sentences* × 10 *sessions* × 2 *emotions* × 8 *speakers*) utterances are used to train *passive* emotion models. The best classification performance was achieved by the models with 64 Gaussian components, converged after 30 iterations. The developed broad emotion models are tested with 300 (5 *sentences* × 10 *sessions* × 3 *emotions* × 2 *speakers*) utterances of active and normal emotions, 200 (5 *sentences* × 10 *sessions* × 2 *emotions* × 2 *speakers*) utterances of passive emotions. The emotion recognition performance on broad emotion groups is shown in Table 5.3. It is evident from the table that the emotion recognition performance for active (fast) and normal emotions is better than that of passive (slow) emotions. This may be because of dominant energy and pitch features for active and normal emotions.

In this work, the weighted score level fusion technique has been adapted for combining the measures from spectral and prosodic features. For each utterance, spectral and prosodic features are extracted separately and individual emotion recognition models are developed. For each feature vector of the test utterance, each model gives the probability that it comes from the respective model. The sum of these vectorwise probabilities gives the effective score for the utterance, corresponding to that model. The model that gives the highest effective score for the utterance is treated as the hypothesized emotion. In the combined approach, the weighted sum of these two effective scores is used to take the decision related to the emotion category. The combination of weights 0.7 and 0.3 respectively for spectral and prosodic features is found to perform better, than the other combinations. The fusion process of combining the measures from spectral and prosodic parameters is shown in Fig. 4.1. After combining the measures, active or fast emotions are recognized without any error. The average recognition performance of the three categories is about 99 %. It is promising performance to continue the study further.

Table 5.4 Finer level classification of active emotions using spectral and prosodic features

Active emotions	Emotion recognition performance		
	Anger	Disgust	Fear
Anger	70	30	00
Disgust	18	82	00
Fear	03	00	97

Table 5.5 Finer level classification of normal emotions using spectral and prosodic features

Normal emotions	Emotion recognition performance		
	Happiness	Neutral	Surprise
Happiness	90	10	00
Neutral	00	100	00
Surprise	20	00	80

Table 5.6 Finer level classification of passive emotions using spectral and prosodic features

Passive emotions	Emo. reco. perf.	
	Sadness	Sarcasm
Sadness	97	03
Sarcasm	00	100

5.5 Finer Level Emotion Recognition

Based on speaking rate, eight emotions of IITKGP-SESC are categorised into 3 broad groups: active, normal and passive emotions in the first stage. Each group in turn contains 2–3 emotions. The classification of these emotions within a group, is referred to as finer level classification. Detailed individual emotion classification performance within each broad emotion category is given in Tables 5.4, 5.5 and 5.6 using the combined evidence of spectral and prosodic features.

From Tables 5.4, 5.5 and 5.6, it may be concluded that, anger, disgust and surprise emotions are not recognized well compared to other emotions. From the subjective evaluation results of IITKGP-SESC (See Table 2.1), conducted to evaluate the quality of the emotion expression, it is observed that the quality of expression of anger, disgust and surprise is not as discriminative as that of other emotions. Table 5.7 shows the comparison of emotion recognition performance using single and two stage classification approaches. From Table, it may be observed that almost all emotions are correctly recognized after the first stage of emotion grouping. In the second stage, around 90 % of average recognition performance is achieved. The emotion recognition performance is observed to be around 76 % using single stage emotion classification. Both two stage and single stage ERSs are developed using the combination of spectral (21 LPCCs) and prosodic (local) features. There is a drastic

Table 5.7 Average emotion classification performance using single and two stage classification approaches on IITKGP-SESC

Emotion groups	Emotions	Emo.reco. after 1st stage	Emo.reco. after 2nd stage	Single stage emo.reco.
Active	Anger	100	70	70
	Disgust		82	70
	Fear		97	77
Normal	Happiness	98	90	83
	Neutral		100	80
	Surprise		80	70
Passive	Sadness	99	97	83
	Sarcasm		100	77
Average		99	89.5	76.25

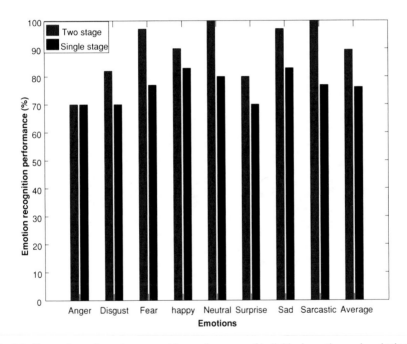

Fig. 5.4 Comparison of emotion recognition performance of individual emotions using single and two stage emotion classification approaches

improvement of about 14 % in the recognition performance, when the two stage approach is employed. This may be visualized in the bar graph shown in Fig. 5.4.

5.6 Summary

In this chapter, to resolve the classification ambiguity of highly confusing emotions, a two-stage classification approach has been proposed to enhance the emotion recognition performance. In this two-stage approach, the combination of spectral and prosodic features has been employed. In the first stage, eight emotions are classified into three broader categories namely active, normal and passive based on the speaking rate. In the second stage, within a broad group, emotions are classified into the individual category. It has been observed that, after the first stage, emotion classification performance is very high. The proposed two stage classification has considerably improved the emotion recognition performance. This method demonstrated the multi-stage emotion classification approach with feature combination.

References

1. S.G. Koolagudi, K.S. Rao, Two stage emotion recognition based on speaking rate. Int. J. Speech Technol. **14**, 35–48 (2011)
2. S.G. Koolagudi, S. Ray, K.S. Rao, Emotion classification based on speaking rate, in *Communications in Computer and Information Science*, ed. by S. Ranka, A. Banerjee, K.K. Biswas, S. Dua, P. Mishra, R. Moona, S.-H. Poon, C.-L. Wang. International Conference on Contemporary Computing, vol. 94, pp. 316–327, Springer, USA, 6–8 Aug 2010
3. K.S. Rao, B. Yegnanarayana, Modeling durations of syllables using neural networks. Comput. Speech Lang. **21**, 282–295 (2007)
4. A.L. Francis, H.C. Nusbaum, Paying attention to speaking rate, in *Fourth International Conference on Spoken Language, 1996 ICSLP 96*, (Philadelphia, PA, USA), pp. 1537–1540 (V3), IEEE, October 1996. Center for Computational Psychology, Department of Psychology, The University of Chicago
5. J. Yuan, M. Liberman, C. Cieri, Towards an integrated understanding of speaking rate in conversation, in *Interspeech 2006*, (Pittsburgh, PA, 2006), pp. 541–544
6. M.S.H. Reddy, K.S. Kumar, S. Guruprasad, B. Yegnanarayana, Subsegmental features for analysis of speech at different speaking rates, in *International Conference on Natural Language Processing*, (Macmillan, India, 2009), pp. 75–80
7. A. LI, Y. ZU, Speaking rate effects on discourse prosody in standard chinese, in *Fourth International Conference on Speech Prosody*, (Campinas, Brazil, 2008), pp. 449–452, 6–9 May 2008
8. H. Yang, W. Guo, Q. Liang, A speaking rate adjustable digital speech repeater for listening comprehension in second-language learning, in *International Conference on Computer Science and, Software Engineering*, vol. 5, pp. 893–896, 12–14 Dec 2008
9. S.G. Koolagudi, S. Maity, V.A. Kumar, S. Chakrabarti, K.S. Rao, IITKGP-SESC : speech database for emotion analysis. Communications in Computer and Information Science, JIIT University, Noida, India: Springer, ISSN: 1865–0929 ed., 17–19 Aug 2009
10. E.F. Lussier, N. Morgan, Effects of speaking rate and word frequency on pronunciations in convertional speech. Speech Commun. **29**, 137–158 (1999)
11. M. Richardson, M.Y. Hwang, A. Acero, X. Huang, Improvements on speech recognition for fast talkers, in *Eurospeech Conference*, Sept 1999

Chapter 6
Emotion Recognition on Real Life Emotions

Abstract Collecting and modelling real life emotions is a real challenge. However, final aim of any emotion recognition system is to identify real world emotions with reasonable accuracy. From the literature it is observed that combination of different features improves the classification performance. In this chapter score level combination of different features has been studied for recognizing real life emotions. For modelling real life emotions, there is a need of good database containing wide variety of real life emotions. In this chapter, Hindi movie database has been used to represent real world emotions. Single and multi-speaker data is collected to study the speaker influence on emotion recognition. Different features are explored for identifying the collected emotions. From the results, it is observed that spectral features carry robust emotion specific information.

6.1 Introduction

In the previous chapters, spectral, prosodic features and their combinations are introduced for robust emotion recognition. Multi-stage emotion classification was also discussed in Chap. 5. However, the research directions are toward recognizing real life emotions. Today's emotion recognition systems recognize studio recorded, full blown emotions with considerably higher accuracy. This data is normally recorded in the constrained environment where there is no background noise, blending of emotions, cross talks. Mostly these full blown emotion databases are recorded using the same text prompts, spoken by the same group of people. Hence, the performance of emotion recognition systems modelled using these databases have a clear influence of speaker and phonetic information. The real challenge of processing the real life emotions is the presence of background music, blended emotions, cross talks and so on. This chapter explores the combination of speech features, for recognizing the real-life emotions. Five emotions collected from Hindi commercial and art movies are used to represent real-life emotions.

K. S. Rao and S. G. Koolagudi, *Robust Emotion Recognition using Spectral and Prosodic Features*, SpringerBriefs in Speech Technology,
DOI: 10.1007/978-1-4614-6360-3_6, © The Author(s) 2013

The aim of any speech emotion recognition system must be to achieve high emotion recognition performance in case of real-life emotions. For this task, a database that represents real-life emotions is an essential component. In this work, real-life emotions are collected from Hindi movies. Spectral features have been explored to classify emotions from real life emotional speech. The reason for choosing the spectral features for this task is their better performance than the source and prosodic features. Gaussian mixture models are used to capture emotion specific information from audio clips of Hindi movies. The detailed results discussed in this work are obtained using 21 LPCC features extracted from the utterance through a conventional block processing approach. At the end of this chapter, the combination of measures from ES, VT system, and local prosodic features is explored on IITKGP-MESC. This helps to analyze the effect of the proposed combination of features on the recognition performance of real-life emotions [1, 2].

6.2 Real Life Emotion Speech Corpus

Practically, it is difficult to a collect real-life emotional speech corpus with a wide variety of emotions. Collecting such emotional data from public and private places leads to the problems related to copyright and privacy issues. Therefore, we have collected real life-like speech emotions from Hindi movies. The emotions expressed in movies are more realistic and contain sufficient linguistic and contextual information. Audiences can easily recognize them. Therefore, IITKGP-MESC (Indian Institute of Technology, Movie Emotion Speech Corpus) is used to represent real-life emotions. The database contains single speaker and multi-speaker speech. Anger, fear, happiness, neutral and sadness are the five emotions collected from Hindi movie clips.

For single speaker speech, video clips of different Hindi movies acted by the same artist are collected separately for different emotions. Later audio tracks from these video clips are separated to obtain the required speech data. This movie speech data is manually preprocessed for removal of non-speech regions like music, silence, and multi-speaker speech that includes cross talks. Fifteen minutes of effective speech data is collected for each emotion from one male and one female actor.

For Multi-speaker emotional speech, the video clips of different Hindi movies are chosen irrespective of actors, for each of the emotions. The emotions from male and female actors are separately collected. The number of actors recorded for each multi-speaker emotion is given in Table 6.1.

The quality of the collected database is evaluated by the subjective listing tests. Five speech segments of 2 s each, in each emotion, are selected and stored as separate files. These 25 files (5 $emotions \times 5\ segments = 25\ files$) are randomly played to the group of 25 Hindi known graduate students of Indian Institute of Technology Kharagpur, India. They are asked to categorize the speech segments into one of the five emotions. The results of subjective emotion classification are shown in Table 6.2. An average of 82 % emotion recognition is achieved by humans. In this study, the

Table 6.1 Number of speakers contributed to each multi-speaker emotion in IITKGP-MESC

Emotion	No. of speakers (female)	No. of speakers (male)
Anger	18	34
Fear	26	35
Happiness	15	12
Neutral	16	12
Sadness	12	15

Table 6.2 Average emotion recognition performance based on subjective listening tests, using IITKGP-MESC

Emotion	Single speaker		Multispeaker	
	(Male)	(Female)	(Male)	(Female)
Anger	92	85	84	82
Fear	75	78	70	73
Happiness	83	89	73	76
Neutral	84	90	78	86
Sadness	90	94	80	89

results are obtained using the multi-speaker Hindi movie emotional speech database. Out of 15 min of data, 12 min of speech data are used for developing the models and the remaining 3 min of data are used for testing. The audio movie clips are divided into the utterances of size 2–3 s for testing.

6.3 Recognition Performance on Real Life Emotions

Emotion recognition using pattern classifiers is basically a two stage process as shown in Fig. 6.1. In the first stage emotion recognition models are developed by training the models using the feature vectors extracted from speech utterances of known emotions. It is known as supervised learning. In the second stage, testing (evaluation) of the trained models is performed by using the speech utterances of unknown emotions. The features of unknown speech are given to all trained models. Then the models compute the probability of unknown feature vectors belonging to the specific model. The model that gives the highest probability is treated as the hypothesized emotion for that feature vector.

The results presented in Table 6.3 are obtained using spectral features of IITKGP-SESC, Emo-DB, and IITKGP-MESC. Spectral features are extracted from an entire utterance, vowel, consonant, and CV transition regions. Pitch synchronous analysis is also done on the speech signal, to extract spectral features from every pitch cycle. Detailed explanation of feature extraction and model building using GMMs

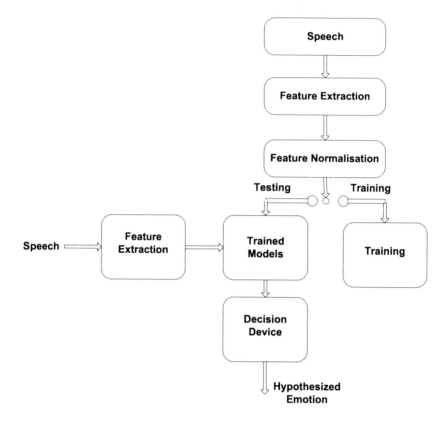

Fig. 6.1   Training and testing phases of developing emotion recognition systems

is given in Chap. 2. The performance of ERSs on IITKGP-MESC is shown in the last column of Table 6.3. The combination of LPCC and formant features has performed quite well for the real life-like emotion recognition task. Real life emotion recognition performance is highest in case of pitch synchronously extracted spectral features. The trend of results is almost the same as in the case of the other two databases. Analyzing the results of three given speech corpora, the emotion recognition performance is considerably high in case of real-life emotions, mostly because the emotions are expressed with proper discrimination, as the movie audio clips contain sufficient linguistic and contextual information.

Real-life emotions are further investigated using other speech features. Emotion recognition studies are conducted on IITKGP-MESC using source, system, and prosodic features. Table 6.4 shows the emotion recognition performance using source, system, and prosodic features individually and in combination. From the results presented in the Table 6.4, it may be observed that there is a considerable improvement in the emotion recognition performance for real life emotions using spectral features, compared to their source and prosodic counterparts. However, a

Table 6.3 Average emotion classification performance using IITKGP-SESC, Emo-DB, and IITKGP-MESC

Features used for developing ERSs	Methods & regions for feature extr.	IIT KGP-SESC	Emo-DB	IIT KGP-MESC
LPCCs	Block	69	64	81
MFCCs	processing	65	63	77
Formant features	(Entire	47	41	53
LPCCs+formants	speech	70	68	82
MFCCs+formants	Signal)	67	63	80
LPCCs	Vowel	54	51	66
MFCCs	region	51	50	62
LPCCs+formants		60	57	70
MFCCs+formants		51	49	61
LPCCs	Consonant	44	40	52
MFCCs	region	44	42	50
LPCCs+formants		51	51	60
MFCCs+formants		36	34	45
LPCCs	CV−	65	66	72
MFCCs	transition	58	57	71
LPCCs+formants	region	71	69	82
MFCCs+formants		68	64	80
LPCCs	Pitch	69	67	83
MFCCs	synchronous	68	65	80
LPCCs+formants	analysis	73	7	84
MFCCs+formants		72	70	81

Table 6.4 Emotion classification performance on IITKGP-MESC using the combination of excitation source, spectral, and prosodic features

Emotions	Emotion recognition performance (%)			
	Excitation source	System features	Prosodic features	Comb. (Spec.+Excit.+Pro.)
Anger	63	83	77	83
Fear	67	80	70	87
Happiness	60	77	67	83
Neutral	57	87	70	90
Sadness	63	80	77	80
Avg.	62.00	81.4	72.20	84.60

marginal improvement in the recognition performance of about 3 % is observed over system features, by using the combination of measures from ES, VT system and prosodic features. Comparison of the results, with respect to each of the emotions present in IITKGP-MESC, is given as a bar graph in Fig. 6.2.

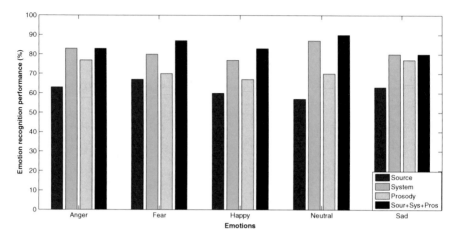

Fig. 6.2 Comparison of emotion recognition performance with respect to each emotion using excitation source, spectral, prosodic, and source+spectral+prosodic features derived on IITKGP-MESC

6.4 Summary

This chapter mainly focussed on emotion recognition on real life emotions. For carrying out this study, speech data was collected from the emotional scenes of various Hindi movies (IITKGP-MESC). Spectral features proposed in Chap. 2 are thoroughly, investigated for discriminating the real life emotions. Emotion recognition performance using spectral features is compared across the emotional databases which represents real life and simulated emotions. In addition to spectral features, excitation source and prosodic features are also analyzed for classifying the real life emotions. Among various speech features, spectral features outperformed the other speech features while recognizing the real life emotions. However, the score-level combination of features have shown the improvement in recognition accuracy.

References

1. S.G. Koolagudi, A. Barthwal, S. Devliyal, K.S. Rao, Real life emotion classification from speech using gaussian mixture models, in *Communications in Computer and Information Science: Contemporary Computing*, ed. by M. Parashar, D. Kaushik, O.F. Rana, R. Samtaney, Y. Yang, A. Zomaya. vol. 306, pp. 250–261, Springer, USA, 6–8 Aug 2012
2. S.G. Koolagudi, S. Devliyal, A. Barthwal, K.S. Rao, Emotion recognition from semi natural speech using artificial neural networks and excitation source features, in *Communications in Computer and Information Science: Contemporary Computing*, ed. by M. Parashar, D. Kaushik, O.F. Rana, R. Samtaney, Y. Yang, A. Zomaya. vol. 306, pp. 273–282, Springer, USA, 6–8 Aug 2012

Chapter 7
Summary and Conclusions

Abstract This chapter summarizes the research work presented in this book, highlights the contributions of the work and discusses the scope for future work. In this book, main attention was given to emotion specific spectral and prosodic features for performing the robust emotion recognition. The book is organized into 7 chapters. The first chapter introduces *speech emotion recognition* as the contemporary research area. In Chap. 2 the spectral features extracted from sub-syllabic regions such as vowels, consonants and CV transition regions are proposed for robust emotion recognition from speech. Pitch synchronously extracted spectral features are also used in Chap. 2 for recognizing the emotions. Chapter 3 proposes use of dynamic prosodic features for recognition of emotions. These dynamic features along with the static prosodic features derived from sentence, word and syllable levels are used for characterizing the emotions. Emotion specific information present in different positions (initial, middle and final) of the speech utterances is used for emotion classification. In Chap. 4, combinations of various emotion specific speech features are explored for developing the robust emotion recognition systems. Chapter 5 deals with the method of multistage emotion classification using combination of features. In this chapter two stage emotion recognition system is developed using spectral and prosodic features. Chapter 6 introduces real life emotion recognition approach using different features. Chapter 7 concludes the present work and flashes light on the directions for further research.

7.1 Summary of the Present Work

Recognizing emotions from speech has emerged as an important research area in the recent past. There are several applications where speech emotion recognition can be deployed. One of the important applications is developing an efficient speech interface to the machine. A properly designed emotional speech database is an essential component for carrying out research on speech emotion recognition. In this work,

K. S. Rao and S. G. Koolagudi, *Robust Emotion Recognition using Spectral and Prosodic Features*, SpringerBriefs in Speech Technology,
DOI: 10.1007/978-1-4614-6360-3_7, © The Author(s) 2013

the simulated emotional speech database in Telugu language (2nd highest spoken language in India) is collected using radio artists in eight common emotions. This database is named as the Indian Institute of Technology KharaGPur- Simulated Emotion Speech Corpus (IITKGP-SESC). The emotions present in the database are anger, disgust, fear, happiness, neutral, sadness, sarcasm, and surprise. The recognition of speech emotions using different speech features is studied using IITKGP-SESC. In this book, mainly three information sources namely (1) Excitation source (2) Vocal tract system and (3) Prosody are explored for robust recognition of speech emotions.

Most of the emotion recognition studies, based on spectral features, have employed conventional block processing, where speech features are derived from the entire speech signal. But, it is known that, the major portion of the speech signal is comprised of steady vowel regions, and there is not much variation in the spectral properties of the signal in this region. Therefore, while extracting spectral features, there is a scope that the features can be extracted from the speech regions that yield spectrally non-redundant information. With this motivation, in this work, spectral features are extracted separately from vowel, consonant and CV transition regions [1]. Similarly, to capture finer spectral variations, pitch synchronous spectral features are used for speech emotion recognition. High amplitude regions of the spectrum are robust while developing speech systems using noisy or real time speech data. In this book, we have used formant features along with other spectral features for recognizing the emotions. In this work GMMs are used for developing the emotion models using spectral features. Among different spectral features, LPCCs seem to perform better in view of discriminating emotions. Recognition performance using formant features is not appreciable, but formant features in combination with other features have shown improvement in the performance. In this work, boundaries of the sub-syllabic segments (consonant, vowel and CV transition) are identified by using vowel onset points. Spectral features of CV transition regions have shown recognition performance close to the performance achieved by the spectral features from the entire utterance. This indicates that the crucial emotion specific information is present in CV transition regions of the syllables. Pitch synchronous spectral features have shown the best recognition performance among various spectral features proposed in this work. This may be due to the finer variations in the spectral characteristics offered by the proposed pitch synchronous spectral features. For the accurate detection of pitch cycles, the zero frequency filter based method is used in this work.

Prosodic features are treated as the effective correlates of speech emotions. In the literature, static prosodic features have been thoroughly investigated for emotion recognition. However, from the perceptual observation of emotional speech, it is observed that emotions are gradually manifested through the sequence of phonemes. This gradual manifestation of emotions may be captured through variations in the articulator movements, while producing emotions. With this motivation, in this book, temporal variations of prosody contours are proposed to capture the emotion specific information. Global and local prosodic features extracted from sentence, word and syllables are explored to discriminate the emotions [2]. The contribution of the above speech segments in different positions (initial, middle, and final) of the sentences and words toward emotion specific knowledge is also systematically studied by developing the emotion recognition models using the features derived from different portions of the speech utterances. In this work, SVM models are explored for capturing the emotion discriminative information from the local and global prosodic features. From the recognition studies using the proposed prosodic features, it is observed that local prosodic features that represent the temporal variation in prosody have more discriminative ability, while classifying the emotions. From the word level prosodic analysis, it was observed that words in final position of the sentence have more emotion discriminative characteristics, and they are almost capable of recognizing emotions at par with sentence level prosodic features. From the syllable level prosodic analysis, it was observed that initial and final syllables have more emotion discriminative capacity than the middle syllables.

The source, system and prosodic features proposed in this book may represent different aspects of speech emotions. Hence, the combination of measures from the proposed supplementary features is investigated to improve the emotion recognition performance of the models. In this work all combinations of the proposed features have enhanced the emotion recognition performance. It indicates that the proposed features represent some non-overlapping emotion specific information. Among various combinations, the highest emotion recognition performance is observed when all three features are combined [3].

A two-stage emotion recognition system has been proposed to improve the emotion recognition performance further. At the first stage emotions are categorized into three broad groups namely, active, passive and normal based on speaking rate. At the second stage, finer classification is performed within each broad group. Here, combinations of spectral and prosodic features are used for developing the emotion models in both stages [4].

Proposed source, system and prosodic features are also explored to recognize real-life natural emotions. Single and multi-speaker real life emotion speech databases are collected from Hindi (Indian national language) movies. Excitation source, spectral and prosodic features are independently, and in combination are used for natural emotion recognition. The robust features presented in this work also perform better in the case of real life emotions [5, 6].

7.2 Contributions of the Present Work

Some of the primary contributions of this book are:

- Design and development of an emotional speech database in Telugu to promote research on speech emotion processing in an Indian context. Design and development of a Hindi movie database to represent real life-like emotions for modeling naturalistic emotions.
- Sub-syllabic and pitch synchronous spectral features are proposed for recognizing the speech emotions.

- Local prosodic features indicating temporal variations in prosodic contours are proposed for recognizing speech emotions. The contribution of different regions (initial, middle, and final) of the utterance toward emotion recognition is analyzed. The global and local prosodic features extracted from sentence, word and syllables are proposed for emotion recognition.
- The combination of measures derived from different speech features is proposed to enhance the performance of emotion recognition systems.
- A two-stage emotion classification system based on speaking rate has been developed to improve the recognition performance over single stage emotion recognition system.
- The proposed source, system, and prosodic features are explored to analyze their discriminative capability with respect to naturalistic emotions (collected from Hindi movies).

7.3 Conclusions from the Present Work

- Spectral features extracted from an entire speech signal may not be essential for classifying the emotions. The features extracted from CV transition regions alone are sufficient to attain appreciably good emotion recognition performance. The emotion recognition systems developed using pitch synchronously extracted spectral features perform slightly better than the systems developed using the features derived from the conventional block processing approach. Out of different spectral features proposed, LPCCs perform better in modeling the emotions.
- Temporal variations in prosodic contours represented by local prosodic features provide more emotion discriminative information than the global prosodic features. In general, final portions of the sentences carry more emotion specific information than the initial and middle portions.
- Combination of measures derived from source, system, and prosodic features has improved the overall emotion recognition performance. This indicates that source, system and prosodic features provide either complementary or supplementary emotion specific information.
- Two stage emotion classification with broad classification of emotions based on speaking rate at the first stage, and finer classification of broad group emotions at the second stage has improved the performance, compared to single stage emotion classification.
- Spectral features extracted from pitch synchronous analysis have shown better recognition performance in the context of real life emotions. Combination of source, spectral and prosodic features has further improved the recognition accuracy.

7.4 Scope for Future Work

- The majority of the research results produced on emotion speech recognition have used databases with limited numbers of speakers. The work discussed in this book also reported the performance based on 10 speakers' speech corpora (IITKGP-SESC and Emo-DB). While developing emotion recognition systems using limited speaker databases, speaker specific information may play a considerable role, if speech utterances of the same speakers are used for training and testing the models. On the other hand, developed models may produce poor results, due to lack of generality, if speech utterances of different speakers are used for training and testing the models. Therefore, there is a need of larger emotional speech databases with reasonably large numbers of speakers and text prompts. Emotion recognition studies have to be conducted on large databases in view of speaker, text and session variabilities.
- This book mainly focuses on characterizing the emotions from a classification point of view. Hence, the main task carried out was deriving the emotion specific information from speech, and using it for classifying the emotions. On the other hand, emotion synthesis through speech is also an important task. Here, emotion specific information may be predicted from the text, and then it has to be incorporated during synthesis. For predicting the emotion specific information, appropriate models have to be developed using a sufficiently large emotion speech corpus. In emotion synthesis, the major issues are design of accurate prediction models and preparation of an appropriate emotion speech corpus.
- Expression of emotions is an universal phenomenon, which may be independent of speaker, gender and language. Cross-lingual emotion recognition study may be another interesting work for further research. The emotion recognition models developed using the utterances of a particular language should yield appreciably good recognition performance for any test utterance of the other language.
- The majority of the work done and results produced in this work are on recognizing speech emotions using a simulated Telugu database. The real challenge is to recognize speech emotions from natural emotions. In this book, the proposed spectral and prosodic features are analyzed on Hindi movie data, which represents the real life emotions. The Hindi movie database mentioned in the book is small and limited. But, for analyzing the realistic and natural emotions, the data should be captured from day-to-day realistic incidents.
- In this book weighted combinations of measures derived from excitation, spectral, and prosodic features are used for improving the emotion recognition performance. The weighting rule used in this work ensures a linear combination of different measures and the sum of the weighting factors is always one. The recognition performance may be further improved by combining this measure with an appropriate non-linear model.
- In this work, a mostly emotion classification task is performed using a single model (i.e., GMM, AANN, or SVM). In future work, hybrid models can be explored for improving the recognition performance. The basic idea behind using the hybrid

models is that they derive the evidence from different perspectives, and hence, the combination of evidence may enhance the performance, if the evidences are supplementary in nature.

- The proposed features and methods in this book are evaluated using a simulated Telugu emotional database and the obtained results are verified on an internationally known Berlin emotion speech database. The trend of emotion recognition is not known on other Indian languages. It would be nice to evaluate the proposed features on different Indian languages for emotion recognition. This helps to comment on whether the methods and features used in this work are language independent. This analysis is also helpful to group the languages based on their emotion characteristics, which in turn would improve the performance of language identification systems.

- In this book, speech emotion recognition is analyzed using different speech features. The study on discrimination of emotions may be extended to the emotion dimensions (arousal, valence and power), which are derived from the psychology of production and perception of emotions. Deriving the appropriate speech features related to the emotion dimensions can be explored for further improving the recognition performance.

- Expression of emotions is a multi-modal activity. Therefore, other modalities like facial expression and bio-signals may be used as the supportive evidence along with the speech signal for developing the robust emotion recognition systems.

- The effect of emotion expression also depends upon the linguistic contents of the speech. Identification of emotion-salient words from the emotional speech, and the features extracted from these words along with other conventional features may enhance the emotion recognition performance.

- In real-time applications such as call analysis in emergency services like ambulance and fire brigade, verification of emotions to analyze the genuineness of the requests is important. In this context, under the framework of emotion verification appropriate features and models can be explored.

- While extracting emotion specific information from the epoch parameters, the accuracy of epoch locations play an important role. The existing epoch extraction methods perform well in the case of clean speech. Therefore, there is a need to strengthen the existing epoch extraction methods to suit realistic situations of speech recording, where recorded speech may contain different background environments. Application of an improved epoch extraction method may further improve the emotion recognition performance.

- In this work, for marking the boundaries of sub-syllabic segments, fixed durations are assigned for vowel, consonant, and CV transition regions with respect to the VOPs. Sometimes it may lead to the overlap of sub-syllabic regions of successive syllables. This may be avoided by marking the boundaries of sub-syllabic regions according to the characteristics of sound units, with respect to the VOPs. With this dynamic segmentation criterion, the emotion recognition performance may be further improved.

- While computing formant features, a simple peak picking algorithm is used to pick up the four dominant peaks from the spectrum. It is observed that sometimes spurious peaks are detected as the formants. Using an efficient formant detection technique may improve emotion recognition performance further.

- Most of today's emotion recognition systems experience a high influence of speaker specific information during emotion classification. In the present work, we normalized the features to minimize speaker dependent information. This is not proved to be an efficient method to nullify speaker specific information. An efficient technique may be developed to remove speaker specific information from the speech utterances.

References

1. S.G. Koolagudi, K.S. Rao, Emotion recognition from speech using sub-syllabic and pitch synchronous spectral features. Int. J. Speech Technol. **15**, 495–511, (Springer, Dec 2012). doi:10.1007/s10772-012-9150-8
2. K.S. Rao, S.G. Koolagudi, R. R. Vempada, Emotion recognition from speech using global and local prosodic features. Int. J. Speech Technol. **15**, (Springer, Aug 2012). doi:10.1007/s10772-012-9172-2
3. S.G. Koolagudi, K.S. Rao, Emotion recognition from speech using source, system and prosodic features. Int. J. Speech Technol. **15**(3), 265–289, (Springer , 2012)
4. S.G. Koolagudi, K.S. Rao, Two stage emotion recognition based on speaking rate. Int. J. Speech Technol. **14**, 35–48 (Mar 2011)
5. S.G. Koolagudi, A. Barthwal, S. Devliyal, K.S. Rao, Real life emotion classification from speech using gaussian mixture models, in *Communications in Computer and Information Science: Contemporary Computing*, ed. by M. Parashar, D. Kaushik, O. F. Rana, R. Samtaney, Y. Yang, A. Zomaya, vol. 306 (Springer, New York, Aug 6–8 2012) pp. 250–261
6. S.G. Koolagudi, S. Devliyal, A. Barthwal, K.S. Rao, Emotion recognition from semi natural speech using artificial neural networks and excitation source features, in *Communications in Computer and Information Science: Contemporary Computing*, ed. by M. Parashar, D. Kaushik, O. F. Rana, R. Samtaney, Y. Yang, A. Zomaya, vol. 306 (Springer, New York, Aug 6–8 2012), pp. 273–282

Appendix A
MFCC Features

The MFCC feature extraction technique basically includes windowing the signal, applying the DFT, taking the log of the magnitude and then warping the frequencies on a Mel scale, followed by applying the inverse DCT. The detailed description of various steps involved in the MFCC feature extraction is explained below.

1. **Pre-emphasis**: Pre-emphasis refers to filtering that emphasizes the higher frequencies. Its purpose is to balance the spectrum of voiced sounds that have a steep roll-off in the high frequency region. For voiced sounds, the glottal source has an approximately -12 dB/octave slope [1]. However, when the acoustic energy radiates from the lips, this causes a roughly $+6$ dB/octave boost to the spectrum. As a result, a speech signal when recorded with a microphone from a distance has approximately a -6 dB/octave slope downward compared to the true spectrum of the vocal tract. Therefore, pre-emphasis removes some of the glottal effects from the vocal tract parameters. The most commonly used pre-emphasis filter is given by the following transfer function

$$H(z) = 1 - bz^{-1} \qquad (A.1)$$

   where the value of b controls the slope of the filter and is usually between 0.4 and 1.0 [1].

2. **Frame blocking and windowing**: The speech signal is a slowly time-varying or quasi-stationary signal. For stable acoustic characteristics, speech needs to be examined over a sufficiently short period of time. Therefore, speech analysis must always be carried out on short segments across which the speech signal is assumed to be stationary. Short-term spectral measurements are typically carried out over 20 ms windows, and advanced every 10 ms [2, 3]. Advancing the time window every 10 ms enables the temporal characteristics of individual speech sounds to be tracked and the 20 ms analysis window is usually sufficient to provide good spectral resolution of these sounds, and at the same time short enough to resolve significant temporal characteristics. The purpose of the overlapping analysis is that each speech sound of the input sequence would be

K. S. Rao and S. G. Koolagudi, *Robust Emotion Recognition using Spectral and Prosodic Features*, SpringerBriefs in Speech Technology, DOI: 10.1007/978-1-4614-6360-3, © The Author(s) 2013

approximately centered at some frame. On each frame a window is applied to taper the signal towards the frame boundaries. Generally, Hanning or Hamming windows are used [1]. This is done to enhance the harmonics, smooth the edges and to reduce the edge effect while taking the DFT on the signal.

3. **DFT spectrum**: Each windowed frame is converted into magnitude spectrum by applying DFT.

$$X(k) = \sum_{n=0}^{N-1} x(n)e^{\frac{-j2\pi nk}{N}} ; \quad 0 \leq k \leq N-1 \tag{A.2}$$

where N is the number of points used to compute the DFT.

4. **Mel-spectrum**: Mel-Spectrum is computed by passing the Fourier transformed signal through a set of band-pass filters known as mel-filter bank. A mel is a unit of measure based on the human ears perceived frequency. It does not correspond linearly to the physical frequency of the tone, as the human auditory system apparently does not perceive pitch linearly. The mel scale is approximately a linear frequency spacing below 1 kHz, and a logarithmic spacing above 1 kHz [4]. The approximation of mel from physical frequency can be expressed as

$$f_{mel} = 2595 \log_{10}\left(1 + \frac{f}{700}\right) \tag{A.3}$$

where f denotes the physical frequency in Hz, and f_{mel} denotes the perceived frequency [2].

Filter banks can be implemented in both time domain and frequency domain. For MFCC computation, filter banks are generally implemented in frequency domain. The center frequencies of the filters are normally evenly spaced on the frequency axis. However, in order to mimic the human ears perception, the warped axis according to the non-linear function given in Eq. (A.3), is implemented. The most commonly used filter shaper is triangular, and in some cases the Hanning filter can be found [1]. The triangular filter banks with mel-frequency warping is given in Fig. A.1.

The mel spectrum of the magnitude spectrum $X(k)$ is computed by multiplying the magnitude spectrum by each of the of the triangular mel weighting filters.

$$s(m) = \sum_{k=0}^{N-1} \left[|X(k)|^2 H_m(k)\right]; \quad 0 \leq m \leq M-1 \tag{A.4}$$

where M is total number of triangular mel weighting filters [5, 6]. $H_m(k)$ is the weight given to the kth energy spectrum bin contributing to the mth output band and is expressed as :

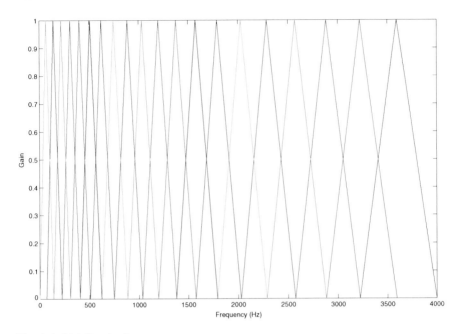

Fig. A.1 Mel-filter bank

$$
H_m(k) = \begin{cases}
0, & k < f(m-1) \\
\frac{2(k-f(m-1))}{f(m)-f(m-1)}, & f(m-1) \le k \le f(m) \\
\frac{2(f(m+1)-k)}{f(m+1)-f(m)}, & f(m) < k \le f(m+1) \\
0, & k > f(m+1)
\end{cases}
\tag{A.5}
$$

with m ranging from 0 to $M-1$.

5. **Discrete Cosine Transform (DCT)**: Since the vocal tract is smooth, the energy levels in adjacent bands tend to be correlated. The DCT is applied to the transformed mel frequency coefficients produces a set of cepstral coefficients. Prior to computing DCT the mel spectrum is usually represented on a log scale. This results in a signal in the cepstral domain with a que-frequency peak corresponding to the pitch of the signal and a number of formants representing low que-frequency peaks. Since most of the signal information is represented by the first few MFCC coefficients, the system can be made robust by extracting only those coefficients ignoring or truncating higher order DCT components [1]. Finally, MFCC is calculated as [1]

$$
c(n) = \sum_{m=0}^{M-1} \log_{10}(s(m)) \cos\left(\frac{\pi n(m-0.5)}{M}\right); \quad n = 0, 1, 2, ..., C-1
\tag{A.6}
$$

where $c(n)$ are the cepstral coefficients and C is the number of MFCCs. Traditional MFCC systems use only 8–13 cepstral coefficients. The zeroth coefficient is often excluded since it represents the average log-energy of the input signal, which only carries little speaker-specific information.

6. **Dynamic MFCC features**: The cepstral coefficients are usually referred to as static features, since they only contain information from a given frame. The extra information about the temporal dynamics of the signal is obtained by computing first and second derivatives of cepstral coefficients [7, 8]. The first order derivative is called delta coefficients, and the second order derivative is called delta-delta coefficients. Delta coefficients tell about the speech rate, and delta-delta coefficients provide information similar to acceleration of speech. The commonly used definition for computing dynamic parameter is

$$\Delta c_m(n) = \frac{\sum\limits_{i=-T}^{T} k_i c_m(n+i)}{\sum\limits_{i=-T}^{T} |i|} \tag{A.7}$$

where $c_m(n)$ denotes the mth feature for the nth time frame, k_i is the ith weight and T is the number of successive frames used for computation. Generally T is taken as 2. The delta-delta coefficients are computed by taking the first order derivative of the delta coefficients.

Appendix B
Gaussian Mixture Model (GMM)

In the speech and speaker recognition the acoustic events are usually modeled by Gaussian probability density functions (PDFs), described by the mean vector and the covariance matrix. However uni-model PDF with only one mean and covariance are unsuitable to model all variations of a single event in speech signals. Therefore, a mixture of single densities is used to model the complex structure of the density probability. For a D-dimensional feature vector denoted as x_t, the mixture density for speaker Ω is defined as weighted sum of M component Gaussian densities as given by the following [9]

$$P(x_t|\Omega) = \sum_{i=1}^{M} w_i P_i(x_t) \tag{B.1}$$

where w_i are the weights and $P_i(x_t)$ are the component densities. Each component density is a D-variate Gaussian function of the form

$$P_i(x_t) = \frac{1}{(2\pi)^{D/2}|\Sigma_i|^{\frac{1}{2}}} e^{-\frac{1}{2}\left[(x_t - \mu_i)'\Sigma_i^{-1}(x_t - \mu_i)\right]} \tag{B.2}$$

where μ_i is a mean vector and Σ_i covariance matrix for ith component. The mixture weights have to satisfy the constraint [9]

$$\sum_{i=1}^{M} w_i = 1. \tag{B.3}$$

The complete Gaussian mixture density is parameterized by the mean vector, the covariance matrix and the mixture weight from all component densities. These parameters are collectively represented by

$$\Omega = \{w_i, \mu_i, \Sigma_i\}; \quad i = 1, 2, ...M. \tag{B.4}$$

K. S. Rao and S. G. Koolagudi, *Robust Emotion Recognition using Spectral and Prosodic Features*, SpringerBriefs in Speech Technology, DOI: 10.1007/978-1-4614-6360-3, © The Author(s) 2013

B.1 Training the GMMs

To determine the model parameters of GMM of the speaker, the GMM has to be trained. In the training process, the maximum likelihood (ML) procedure is adopted to estimate model parameters. For a sequence of training vectors $X = \{x_1, x_2, .., x_T\}$, the GMM likelihood can be written as (assuming observations independence) [9]

$$P(X|\Omega) = \prod_{t=1}^{T} P(x_t|\Omega). \qquad (B.5)$$

Usually this is done by taking the logarithm and is commonly named as log-likelihood function. From Eqs. (B.1) and (B.5), the log-likelihood function can be written as

$$\log[P(X|\Omega)] = \sum_{t=1}^{T} \log\left[\sum_{i=1}^{M} w_i P_i(x_t)\right]. \qquad (B.6)$$

Often, the average log-likelihood value is used by dividing $\log[P(X|\Omega)]$ by T. This is done to normalize out duration effects from the log-likelihood value. Also, since the incorrect assumption of independence is underestimating the actual likelihood value with dependencies, scaling by T can be considered a rough compensation factor [10]. The parameters of a GMM model can be estimated using maximum likelihood (ML) estimation. The main objective of the ML estimation is to derive the optimum model parameters that can maximize the likelihood of GMM. The likelihood value is, however, a highly nonlinear function in the model parameters and direct maximization is not possible. Instead, maximization is done through iterative procedures. Of the many techniques developed to maximize the likelihood value, the most popular is the iterative expectation maximization (EM) algorithm [11].

B.1.1 Expectation Maximization (EM) Algorithm

The EM algorithm begins with an initial model Ω and tends to estimate a new model such that the likelihood of the model increasing with each iteration. This new model is considered to be an initial model in the next iteration and the entire process is repeated until a certain convergence threshold is obtained or a certain predetermined number of iterations have been made. A summary of the various steps followed in the EM algorithm are described below.

1. **Initialization**: In this step an initial estimate of the parameters is obtained. The performance of the EM algorithm depends on this initialization. Generally,

LBG [12] or K-means algorithm [13, 14] is used to initialize the GMM parameters.

2. **Likelihood Computation**: In each iteration the posterior probabilities for the ith mixture is computed as [9]:

$$\Pr(i|x_t) = \frac{w_i P_i(x_t)}{\sum_{j=1}^{M} w_j P_j(x_t)}. \tag{B.7}$$

3. **Parameter Update**: Having the posterior probabilities, the model parameters are updated according to the following expressions [9].
Mixture weight update:

$$\overline{w_i} = \frac{\sum_{i=1}^{T} \Pr(i|x_t)}{T}. \tag{B.8}$$

Mean vector update:

$$\overline{\mu_i} = \frac{\sum_{i=1}^{T} \Pr(i|x_t)x_t}{\sum_{i=1}^{T} \Pr(i|x_t)}. \tag{B.9}$$

Covariance matrix update:

$$\overline{\sigma_i^2} = \frac{\sum_{i=1}^{T} \Pr(i|x_t)|x_t - \overline{\mu_i}|^2}{\sum_{i=1}^{T} \Pr(i|x_t)}. \tag{B.10}$$

In the estimation of the model parameters, it is possible to choose, either full covariance matrices or diagonal covariance matrices. It is more common to use diagonal covariance matrices for GMM, since linear combination of diagonal covariance Gausses has the same model capability with full matrices [15]. Another reason is that speech utterances are usually parameterized with cepstral features. Cepstral features are more compatible, discriminative, and most important, they are nearly uncorrelated, which allows diagonal covariance to be used by the GMMs [9, 16]. The iterative process is normally carried out 10 times, at which point the model is assumed to converge to a local maximum [9].

B.1.2 *Maximum A Posteriori (MAP) Adaptation*

Gaussian mixture models for a speaker can be trained using the modeling described earlier. For this, it is necessary that sufficient training data is available in order to create a model of the speaker. Another way of estimating a statistical model, which is especially useful when the training data available is of short duration, is by using maximum *a posteriori* adaptation (MAP) of a background model trained on the speech data of several other speakers [17]. This background model is a large GMM that is trained with a large amount of data which encompasses the different kinds of speech that may be encountered by the system during training. These different kinds may include different channel conditions, composition of speakers, acoustic conditions, etc. A summary of MAP adaptation steps are given below.

For each mixture i from the background model, $Pr(i|x_t)$ is calculated as [18]

$$Pr(i|x_t) = \frac{w_i P_i(x_t)}{\sum\limits_{j=1}^{M} w_j P_j(x_t)}.$$ (B.11)

Using $Pr(i|x_t)$, the statistics of the weight, mean and variance are calculated as follows [18]

$$n_i = \sum_{i=1}^{T} Pr(i|x_t)$$ (B.12)

$$E_i(x_t) = \frac{\sum\limits_{i=1}^{T} Pr(i|x_t)x_t}{n_i}$$ (B.13)

$$E_i(x_t^2) = \frac{\sum\limits_{i=1}^{T} Pr(i|x_t)x_t^2}{n_i}.$$ (B.14)

These new statistics calculated from the training data are then used adapt the background model, and the new weights ($\hat{w}_i$), means ($\hat{\mu}_i$) and variances ($\hat{\sigma}_i^2$) are given by [18]

$$\hat{w}_i = \left[\frac{\alpha_i n_i}{T} + (1 - \alpha_i)w_i \right] \gamma$$ (B.15)

$$\hat{\mu}_i = \alpha_i E_i(x_t) + (1 - \alpha_i)\mu_i$$ (B.16)

$$\hat{\sigma}_i^2 = \alpha_i E_i(x_t^2) + (1 - \alpha_i)(\sigma_i^2 + \mu_i^2) - \hat{\mu}_i^2.$$ (B.17)

A scale factor γ is used, which ensures that all the new mixture weights sum to 1. α_i is the adaptation coefficient which controls the balance between the old and new model parameter estimates. α_i is defined as [18]

$$\alpha_i = \frac{n_i}{n_i + r} \tag{B.18}$$

where r is a fixed relevance factor, which determines the extent of mixing of the old and new estimates of the parameters. Low values for α_i ($\alpha_i \rightarrow 0$), will result in new parameter estimates from the data to be de-emphasized, while higher values ($\alpha_i \rightarrow 1$) will emphasize the use of the new training data-dependent parameters. Generally only mean values are adapted [10]. It is experimentally shown that mean adaptation gives slightly higher performance than adapting all three parameters [18].

B.2 Testing

In identification phase, mixture densities are calculated for every feature vector for all speakers and speaker with maximum likelihood is selected as identified speaker. For example, if S speaker models $\{\Omega_1, \Omega_2,..., \Omega_S\}$ are available after the training, speaker identification can be done based on a new speech data set. First, the sequence of feature vectors $X = \{x_1, x_2, .., x_T\}$ is calculated. Then the speaker model $\hat{s}$ is determined which maximizes the a posteriori probability $P(\Omega_S|X)$. That is, according to the Bayes rule [9]

$$\hat{s} = \max_{1 \leq s \leq S} P(\Omega_S|X) = \max_{1 \leq s \leq S} \frac{P(X|\Omega_S)}{P(X)} P(\Omega_S). \tag{B.19}$$

Assuming equal probability of all speakers and the statistical independence of the observations, the decision rule for the most probable speaker can be redefined as

$$\hat{s} = \max_{1 \leq s \leq S} \sum_{t=1}^{T} \log P(x_t|\Omega_s) \tag{B.20}$$

with T the number of feature vectors of the speech data set under test and $P(x_t|\Omega_s)$ given by Eq. (B.1).

Decision in verification is obtained by comparing the score computed using the model for the claimed speaker Ω_S given by $P(\Omega_S|X)$ to a predefined threshold θ. The claim is accepted if $P(\Omega_S|X) > \theta$, and rejected otherwise [10].

References

1. J.W. Picone, Signal modeling techniques in speech recognition. Proc. IEEE **81**, 1215–1247 (1993)

2. J.R. Deller, J.H. Hansen, J.G. Proakis, *Discrete Time Processing of Speech Signals*, 1st edn. (Prentice Hall PTR, Upper Saddle River, 1993)
3. J. Benesty, M.M. Sondhi, Y.A. Huang, *Springer Handbook of Speech Processing* (Springer, New York, 2008)
4. J. Volkmann, S. Stevens, E. Newman, A scale for the measurement of the psychological magnitude pitch. J. Acoust. Soc. Am. **8**, 185–190 (1937)
5. Z. Fang, Z. Guoliang, S. Zhanjiang, Comparison of different implementations of MFCC. J. Comput. Sci. Technol. **16**(6), 582–589 (2001)
6. G.K.T. Ganchev, N. Fakotakis, Comparative evaluation of various MFCC implementations on the speaker verification task. in *Proceedings of International Conference on Speech and Computer*, Patras, Greece, pp. 191–194 (2005)
7. S. Furui, Comparison of speaker recognition methods using statistical features and dynamic features. IEEE Trans. Acoust. Speech Signal Process. **29**(3), 342–350 (1981)
8. J. S. Mason, X. Zhang, Velocity and acceleration features in speaker recognition, in *Proceedings of the IEEE International Conference on Acoustics, Speech, Signal Processing*, Toronto, Canada, pp. 3673–3676, April 1991
9. D.A. Reynolds, Speaker identification and verification using Gaussian mixture speaker models. Speech Commun. **17**, 91–108 (1995)
10. F. Bimbot, J.F. Bonastre, C. Fredouille, G. Gravier, M.I. Chagnolleau, S. Meignier, T. Merlin, O.J. Garcia, D. Petrovska, D.A. Reynolds, A tutorial on text-independent speaker verification. EURASIP J. Appl. Signal Process. **4**, 430–451 (2004)
11. A. Dempster, N.M. Laird, D.B. Rubin, Maximum likelihood from incomplete data via the EM algorithm. J. R. Stat. Soc. **39**(1), 1–38 (1977)
12. Y. Linde, A. Buzo, R. Gray, An algorithm for vector quantizer design. IEEE Trans. Commun. **28**, 84–95 (1980)
13. J.B. MacQueen, Some methods for classification and analysis of multivariate observations, ed. by L.M.L. Cam, J. Neyman. in *Proceedings of the Fifth Berkeley Symposium on Mathematical Statistics and Probability*, vol. 1 (University of California Press, Berkeley, 1967), pp. 281–297
14. J.A. Hartigan, M.A. Wong, A K-means clustering algorithm. Appl. Stat. **28**(1), 100–108 (1979)
15. Q.Y. Hong, S. Kwong, A discriminative training approach for text-independent speaker recognition. Signal Process. **85**(7), 1449–1463 (2005)
16. D. Reynolds, R. Rose, Robust text-independent speaker identification using Gaussian mixture speaker models. IEEE Trans. Speech Audio Process. **3**, 72–83 (1995)
17. J. Gauvain, C.-H. Lee, Maximum a posteriori estimation for multivariate Gaussian mixture observations of Markov chains. IEEE Trans. Speech Audio Process. **2**, 291–298 (1994)
18. D.A. Reynolds, Speaker verification using adapted Gaussian mixture models. Digital Signal Process. **10**, 19–41 (2000)

Printed by Publishers' Graphics LLC
DBT130506.15.24.72